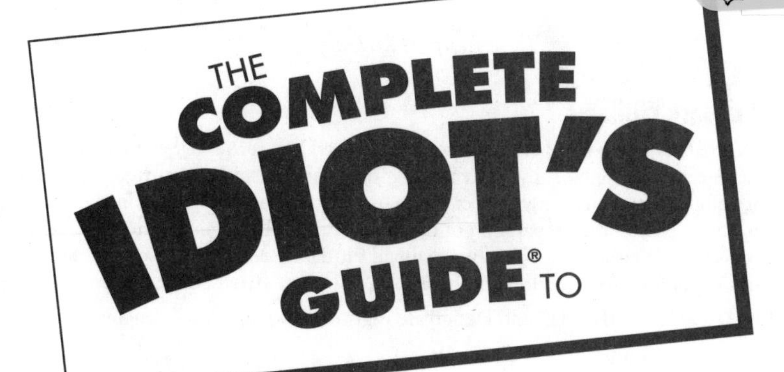

The Arctic and Antarctic

by Jack Williams

ALPHA

A member of Penguin Group (USA) Inc.

To my wife, Darlene.

International Standard Book Number: 1-59257-073-9
Library of Congress Catalog Card Number: 2003105468

05 04 03 8 7 6 5 4 3 2 1

Interpretation of the printing code: The rightmost number of the first series of numbers is the year of the book's printing; the rightmost number of the second series of numbers is the number of the book's printing. For example, a printing code of 03-1 shows that the first printing occurred in 2003.

Printed in the United States of America

Publisher: *Marie Butler-Knight*
Product Manager: *Phil Kitchel*
Senior Managing Editor: *Jennifer Chisholm*
Acquisitions Editor: *Mikal Belicove*
Development Editor: *Tom Stevens*
Copy Editor: *Jeff Rose*
Illustrator: *Chris Eliopoulos*
Cover/Book Designer: *Trina Wurst*
Indexer: *Tonya Heard*
Layout/Proofreading: *Megan Douglass, Becky Harmon*
Graphics: *Tammy Graham, Becky Harmon, Dennis Sheehan, Laura Robbins*

Contents at a Glance

Appendixes

Contents

Appendixes

Foreword

My fascination with Antarctica began when, at age 12, I picked up a copy of Alfred Lansing's *Endurance* from my parents' bookshelf. I was transfixed by Frank Hurley's amazing photos of Shackleton and his men and by their incredible story of survival. I wanted to help chip the ship out of the ice, care for the dogs, and join in those soccer games on the sea ice. Dreams of skiing in polar regions took hold of a young Minnesota girl and never went away.

I have been lucky enough to be able to live out many of those dreams (some might say "crazy enough"; believe me, "lucky" is the right adjective). My explorations of polar regions began when I joined 7 men and 49 male dogs as the only female member of the Steger International Polar Expedition, earning the distinction of being the first known woman in history to cross the ice to the North Pole. In 1993, I led a group of four women on the American Women's Expedition, when we skied from the edge of Antarctica to the South Pole. In 2000–2001, I joined my Norwegian ski partner Liv Arnesen on the Bancroft Arnesen Expedition, on which we became the first women to ski and sail across the Antarctic landmass.

For me, my polar journeys have been much less about "historic firsts" than about the chance to experience firsthand the magic of the polar regions and to share my experiences with others. No book can fully convey that magic. But in *The Complete Idiot's Guide to the Arctic and Antarctic*, Jack Williams gives you glimpses of the wonder. You'll learn fascinating facts about the poles. You'll know why we had to take a rifle with us on our North Pole Expedition, and why a rifle would have just been excess baggage on our Antarctic expeditions. You'll get an understanding of the important role that the polar regions play in weather all around the world. You'll know that glacial ice looks blue. And you'll understand *why* it looks blue.

I hope that somewhere, sometime soon, a 12-year-old girl or boy will pick up *The Complete Idiot's Guide to the Arctic and Antarctic* and begin a love affair with the poles like I did. Or that a 40-year-old will pick up this book and rekindle an interest in polar regions that had "gone cold." This book is a repository of lots of interesting and accurate information about the polar regions. It can also be a great springboard to other sources of information.

Over the last 40-some years, I've learned lots about the Arctic and Antarctic regions. I've picked up information a bit here, a bit there. *The Complete Idiot's Guide to the Arctic and Antarctic* provides an entertaining and efficient way to get lots of information and inspiration. I wish it was available 35 years ago—when I knew very little about the poles and wanted to know everything.

Enjoy!

Ann Bancroft

Ann Bancroft is one of the world's preeminent female polar explorers. Ann is the first woman to cross the ice to both the North and South Poles. In 1986, Ann skied and dogsledded 1,000 miles from the Northwest Territories in Canada to the North Pole as the only female member of the Steger International Polar Expedition. In 1993, she led the American Women's Expedition, a 67-day, 4-woman expedition of 660 miles on skis from the coast of Antarctica to the South Pole. Recently, Ann and Norwegian polar explorer Liv Arnesen became the first woman in history to sail and ski across Antarctica's landmass—completing a 94-day, 1,717-mile trek in 2001. Since the Bancroft Arnesen Expedition, Ann has helped shape YourExpedition.com, a 12-person, international motivational company that offers organizations and individuals inspiration and guidance to succeed in life's expeditions. Her new book, *No Horizon Is So Far,* which she co-wrote with Liv Arnesen, will be available in September 2003 from DaCapo Press.

Introduction

Obviously we don't think you are an "idiot." Instead, we assume you are an intelligent person who doesn't happen to know much about the Arctic and Antarctic. You are probably like most people in that you aren't sure which of the polar regions is mostly ocean and which is mostly land, which one has polar bears, and which has penguins.

We also assume that you might be interested in learning certain things about the Arctic or Antarctic, and that you might want to skip some of the chapters, at least as you begin reading. For instance, you might see on a television news broadcast that polar ice is melting and want to find out what's going on. In this case, you would probably go right to Chapter 27 and see what you can learn. By doing this, you will be able to discover the gist of what's happening to polar ice. As you read the chapter, you will see references to earlier chapters where we explain some of the basic science, and by looking at those chapters, you'll learn more.

No matter what aspect of the polar regions you are interested in, you might want to begin with the five chapters in **Part 1, "Meet the Polar Regions."** They introduce you to the Arctic and Antarctic, their weather, their very different landscapes, the ice that dominates both land and sea in both regions, and finally the cold oceans of the Arctic and Antarctic.

Part 2, "Life in the Polar Regions," introduces you to the plants and animals that live on land and in the sea of both the Arctic and Antarctic.

Part 3, "Going to the Ends of the Earth," shifts to people, beginning with those who settled around the Arctic around 5,000 years ago. In the rest of the section, we follow the adventures and misadventures of those who explored the Arctic and Antarctic through World War II and into the Cold War era.

Part 4, "The Polar Regions Today," begins with a look at the people, politics, and problems of the polar regions today in Chapters 15 and 16. Chapter 17 is about aviation in Antarctica because that has become the main mode of travel there. Chapter 18 tells you about working in Antarctica, what it's like and even how to go about looking for a job there. With Chapter 19, we close Part 4 by helping you find the polar vacation that fits your tastes and budget.

Part 5, "Today's Polar Science," tells what scientists are up to in both the Arctic and Antarctic, how they are going about their work and what they are learning or hope to learn. Both polar regions have been thoroughly explored in the sense of locating the mountains, bays, islands, and glaciers and precisely measuring how far each place is from the others. As you'll see in Part 5, scientists are doing today's real polar exploration.

Finally, **Part 6, "The Polar Regions and the Rest of the World,"** looks at how the polar regions are connected with the rest of the world and how their ice is telling scientists about climates of the past. Then we finish the book by looking at what seems to be happening now and what seems to be likely in the future if the earth's climate continues changing as it has in the last few decades.

Extras

In addition to helping you understand the polar regions with the main text and illustrations, we help break your trail through the polar world with extra tidbits of information.

Ice Chips

These are little bits of information that might give you a chuckle, or add some information that's handy, but not absolutely necessary for understanding the main points of their chapter.

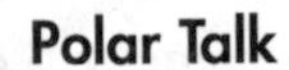

Polar Talk

These definitions tell you what unfamiliar polar terms mean, or in some cases, what some familiar words or terms mean when used by polar explorers or scientists. We also define some scientific terms when needed.

Crevasse Caution

Think of these like the red flags used to mark dangerous areas on the ice, such as crevasses. If you ignore them, you won't fall into a narrow crack in the ice, but you might leave the chapter with a misleading idea about something.

Explorations

These boxes include information that's not absolutely needed to understand the chapter but that gives you more information, or a deeper insight into the topic the chapter covers.

Acknowledgments

No book is the work of an author alone, and many people helped get this book started and kept it going.

I'd like to start with Scott Waxman, my agent, who brought the idea to me, and the Alpha Books editors I worked with, Eric Heagy, Mikal Belicove, and Tom Stevens, who kept their cool when I stumbled. April Holladay, a first-rate, freelance science writer, did the research and first drafts for three of the chapters. Joseph Frey, an accomplished, Canadian science writer, did research and the first drafts of parts of two other chapters.

I couldn't begin to list all of the scientists, research camp managers and workers, and many others in Greenland; Antarctica; Barrow, Alaska; on Coast Guard icebreakers; and with the New York Air National Guard's 109th Airlift Wing, who helped me learn about science, life, work, and travel in the polar regions. Public affairs officers with the National Science Foundation, especially Peter West in recent years, have always been quick to help me find information. Valerie Carroll, public affairs officer for Raytheon Polar Services, has also been very helpful, especially in helping me find photos for this book. Major Bob Bullock of the New York Air National Guard has been a font of polar aviation knowledge over the years.

Finally I must thank my wife, Darlene Shields, who not only carefully edited each chapter as I worked on this book, but who also continues to brighten my life with her loving support. She makes my world a warm place.

Special Thanks to the Technical Reviewer

The Complete Idiot's Guide to the Arctic and Antarctic was reviewed by an expert who double-checked the accuracy of what you'll learn here, to help us ensure that this book gives you everything you need to know about the Arctic and Antarctic. Special thanks are extended to Michael Ledbetter.

Trademarks

Part 1

Meet the Polar Regions

Welcome to the fascinating world of the Arctic and Antarctic—the polar regions. You'll discover that there's more to polar weather than icy cold. You'll discover that the Arctic is mostly ocean, while the Antarctic is mostly land, both characterized by a lot of ice. We'll see what makes ice act the way it does, including why it floats. Finally, we'll conclude our introductory tour by looking at their two icy oceans, which turn out to be a lot more important for the entire world than you might think.

Chapter 1

Meet the Arctic and Antarctic

In This Chapter

- Why the Arctic and Antarctic are fascinating
- More than cold defines the Polar Regions
- Day and night lose their meaning near the poles
- A December visit to the Arctic Coast
- A race around the world at the South Pole

Maybe you picked up this book because you saw the television documentary or the movie about the ice around Antarctica trapping Sir Ernest Shackleton's ship in 1914 and then crushing it. The story of how Shackleton and all the men with him returned home safely might have sparked a desire to learn more about the people who first ventured into Earth's polar regions. Who were they? Why did the go? What did they find? How did they survive? Is going there now as challenging as it was almost a century ago? With global warming so much in the news, you might be curious about what is going on in the Polar Regions today. Is the ice in the Arctic and Antarctic about to melt and drown our coastal cities? Maybe your spirit of adventure is pushing you to learn about the frigid ends of the earth because you want to either find a job there or visit as a tourist.

The Complete Idiot's Guide to the Arctic and Antarctic will answer these and many more questions. In this chapter, we'll introduce you to the Polar Regions with a few glimpses of the Arctic and Antarctic, including the South Pole, which is as far south as you can go. These will give you a taste of what the Polar Regions are like, and a sense of the size and strangeness of the ends of the earth.

Polar Means More Than Cold

The first picture of the *Arctic* or *Antarctic* that's likely to pop into your mind is frigid air, snow, and ice. Although it's true that Antarctica, on the average, is the coldest place in the world, more than cold distinguishes the polar regions from the rest of the world. In some parts of *Antarctica* you'd be comfortable outdoors in a light jacket on some days. For instance, on an average January day at the big U.S. McMurdo Station in Antarctica, the high and low temperatures of 32°F (0°C) and 22°F (–5°C) are close to those in Boston the same month. Although the temperatures are similar, McMurdo and Boston differ in one important way during January: In McMurdo, it's the middle of summer; in Boston, the middle of winter.

Polar Talk

The **Arctic** is the Northern Hemisphere polar region surrounding the North Pole, extending about 1,621 miles (2,608 kilometers) south all around the earth. The **Antarctic** is the Southern Hemisphere polar region surrounding the South Pole, extending about 1,621 miles (2,608 kilometers) north all around the earth. **Antarctica** is the only continent in the Antarctic.

An even more important difference between the two places in January is that if you stayed in McMurdo all month you'd never see the sun set. If you were in Boston early on a January morning, with the sun not coming up until after 7 A.M., you might wonder whether its ever going to rise.

The nature of day and night are the heart of what makes the polar regions different from other parts of the earth. The few people who venture to the North or South poles go to places where the sun doesn't set for six months each year and doesn't rise the other six months. If you travel away from the either pole toward the equator, you would see the sun rise and set on more and more days during the year until you are more than 1,621 miles from the Pole. Here you would see that the sun rises and sets each day of the entire year. We usually consider the polar regions to be the parts of the earth with at least one day a year without a sunrise and another day without a sunset.

Explorations

The Arctic Circle and Antarctic Circle divide parts of the earth that see the sun rise and set every day of the year from parts with at least one day with no sunset and one day with no sunrise. The circles are at approximately 66.5 degrees north and south latitudes. We say "approximately" because the earth wobbles a little on its axis; the lines can move 25 feet or more in a year. You might see a midnight sun a few miles outside the polar regions because the atmosphere bends the sun's light and you see the sun when it's really below the horizon. Inside the polar regions the sun can move behind a hill or mountain, blocking it from view as a sunset does.

Temperatures Plunge on Long Nights

Long polar nights are the main cause of polar cold because the sun is the source of all but a tiny part of the earth's warmth. Even when the sun is in the sky 24 hours a day during the summer, it doesn't melt all of the polar ice because there's so much snow and ice. Warming it and then melting it takes a lot of energy. Also, snow and ice reflect away large amounts of solar energy.

Although all the Arctic has long nights, some places are much warmer than others even during the winter. For instance, warm water from the Gulf Stream and its eastern extension, the North Atlantic Drift, makes Europe's Arctic Ocean Coast much warmer than Alaska's, which is at almost the same latitude. On the other hand, Fairbanks, Alaska, which is south of the Arctic Circle, is often colder on winter days than any place in the state north of the Arctic Circle. Glaciers, which are expanses of ice that form where snow doesn't melt during the summer, are found far from the polar regions, including even on top of tropical mountains.

Dark Cold Makes Living Anything but Easy

Cold is a huge challenge to all forms of life, including humans. On the most simple level, when the water in cells freezes it breaks the cells, causing death. But the effects of cold are even more far reaching. Low temperatures change the basic biochemistry of living things, usually to ill effect. In addition, plants—the bottom of the food web—can't grow when there's no light. Despite these hurdles, life thrives on both land and sea in the Arctic and in the ocean around Antarctica. The miles-thick ice that covers almost all of Antarctica is a barrier to all forms of life except a few microbes. You might be wondering about photos you've seen of colonies of hundreds of penguins around Antarctica. These colonies are on the coast because penguins find all of their food in the ocean. The only Antarctic birds that go to dinner on land are those that eat the eggs of penguins and other birds.

CAUTION

Crevasse Caution

Despite what most world maps show, Greenland is not almost as big as South America and Antarctica isn't a narrow strip. Maps distort the Polar Regions because it's impossible to represent the spherical Earth on a flat map and mapmakers are usually interested in places away from the poles. Use a globe or maps centered on the North and South poles to get a truer picture.

Elsewhere in the polar regions, however, an amazing number of plants and animals, from microscopic algae to polar bears, have evolved ways to survive months of cold darkness. Other animals migrate to warmer places during the winter, including some birds that fly away to find summer in the earth's other polar region.

Humans have lived in the Arctic for around 4,000 years, discovering reliable sources of food that don't depend on growing crops, and developing the clothing and shelter needed to survive the long, cold night. As Europeans began exploring the Arctic and Antarctic they often failed to heed the lessons the Arctic's native people had to teach. Mastering the survival methods of Arctic natives and improving on them with modern technology opened the way for humans to explore not only the Arctic, but also Antarctica, the only continent that has never had a native population.

Let's Take an Arctic Journey

We tend to think of the polar regions as being far away and hard to reach. Although that's true of Antarctica, some parts of the Arctic are quite easy to reach and not all that expensive to get to. In Europe, for example you can take a train to Bodo, Norway, which is about 75 miles north of the Arctic Circle.

In the United States, you can have your photo taken at an "Arctic Circle" sign by driving about 150 miles north from Fairbanks, Alaska. For a better taste of the Arctic, you can fly from Anchorage or Fairbanks to Barrow, the northernmost city in the United States.

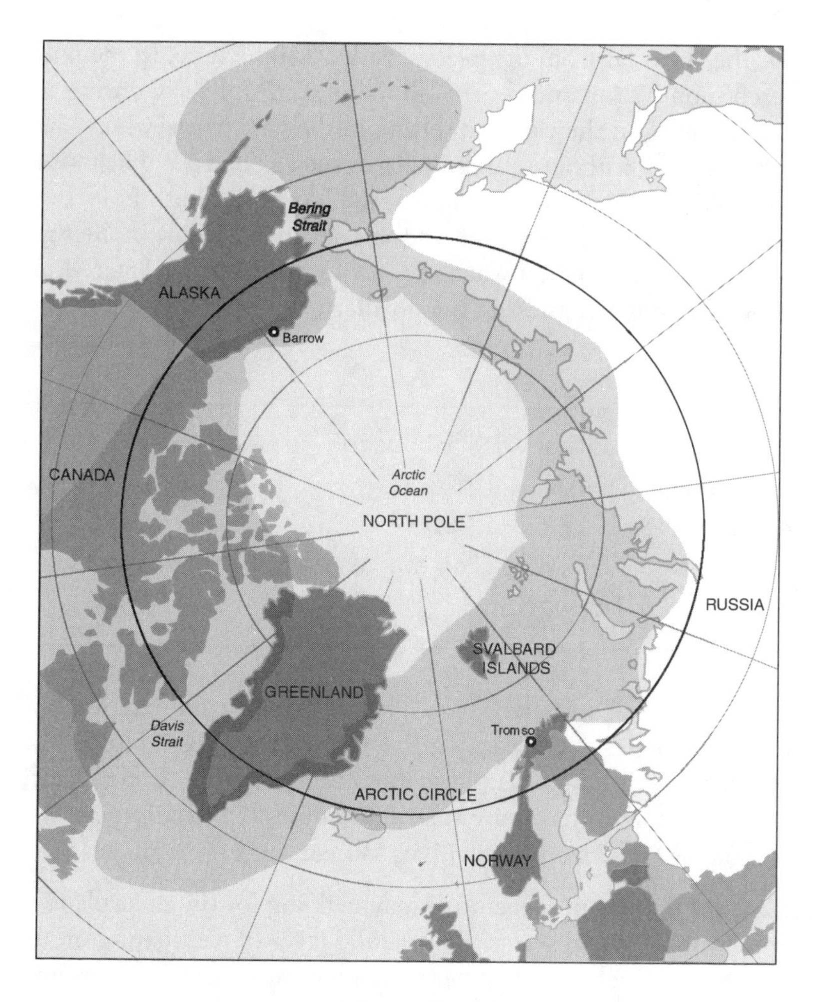

Map of the Arctic.

Winter Sky Isn't Completely Dark

A winter visit to Barrow would be the least-expensive and time-consuming way to get a real taste of just how different the polar regions are from the rest of the world. To begin with, even though the sun sets in Barrow on November 18 and doesn't rise again until January 23, it's not as dark as you might think, even on December 22, the day the sun is the lowest in the Northern Hemisphere sky. On that day, which is the winter solstice, Barrow has *civil twilight* from around noon until 3 P.M. On a cloud-free day it becomes bright enough to move around outside without lights.

Polar Talk

Civil twilight means that the sun's center is less than 6 degrees below the horizon. It's usually light enough to engage in outdoor activities without artificial light.

Even after twilight ends, the sky is far from completely dark all of the time. In the polar regions, for instance, each month's full and nearly full moon circles the sky during winter's darkness instead of rising in the east and setting in the west as elsewhere. In the summer the sun circles the sky without setting in places such as Barrow. The circles the sun and moon make are tilted unless you are right at the North or South Pole. In Barrow, as in other Arctic places away from a Pole, the sun or moon are low in the sky when they circle around to the north, highest when they are in the south. In Antarctica, away from the South Pole, the sun is highest in the northern sky, lowest in the south.

Ice Chips

Highs, lows, and averages of December weather in Barrow, Alaska:

- December average high: –5°F (–21°C)
- December average low: –17°F (–27°C)
- December warmest: 32°F (0°C) (highest ever recorded during the year)
- December coldest: –51°F (–46°C)

With the darkness and cold, a visitor to Barrow could be forgiven for thinking the North Pole is near. One of the places tourists stay is the "Top of the World Hotel." But the top of the world, the North Pole, is far away. If you stand on the Arctic Ocean beach in Barrow, the North Pole is a little more than 1,300 miles (2,092 kilometers) due north. This is just about as far as from Green Bay, Wisconsin, to Miami.

If you could head north from Barrow in December, maybe flying low in an airplane under clear skies and a full moon, you'd see an almost solid layer of ice floating on the ocean, broken here and there by openings exposing the frigid water. If you're lucky, you might spot a hungry polar bear on the ice near an opening, waiting for a seal to come up for air. You might even see a snowy owl. Scientists who have put tiny radio transmitters on snowy owls have discovered that some spend at least part of the winter darkness on the Arctic's ice. The scientists' best guess is that the owls feed on the leftovers from polar bear dinners. The Arctic Ocean isn't as lifeless, even in winter, as its icy visage might make you think.

Ice Chips

The Arctic is a major airline route with jets regularly flying over the region, including over the North Pole. If you look at the top of a globe you'll see why: Going over the Arctic is the shortest route to and from Asia and the United States and Asia and Europe.

It's a Long Way Across

Once you reach the North Pole on an imaginary flight, no matter which way you turn you'll be heading south. If you made a slight turn to the east shortly after passing the pole, and fly something more than 600 miles (965 kilometers), you'd see Norway's Svalbard Islands. Spitsbergen is the largest. Another 600 or so miles across the Norwegian Sea brings you to Tromso on Norway's Arctic Coast. A slight turn to the west after passing the North Pole would take you across the Barents Sea to Murmansk, the Russian Arctic Ocean port.

No matter which of the infinite number of "souths" you take across the 5.5 million square miles of the Arctic Ocean from the North Pole, you are much more likely to hit land than water. The Arctic connects with the Pacific via the 45-mile-wide Bering Strait between Siberia and Alaska. The connections with the Atlantic are much wider—about 950 miles between Greenland and Norway, a little more than 200 miles across the Davis Strait between Greenland and Canada's Baffin Island, and the 90 miles or so across Hudson Strait between Baffin Island and Newfoundland. The fact that the center of the Arctic is an ocean that's connected with the world's other oceans has important consequences not only for the Arctic, but also for the world's weather and *climate*.

Polar Talk

Climate is the average weather for a particular place or region. Meteorologists usually use statistical information averaged over 30 years to give "normal" temperatures, precipitation, and other factors for a location. Information on weather extremes also helps define a place's climate.

Icy Continent Makes Life a Struggle

Antarctica is the earth's most remote place. Most of the tourists who visit go on ships that leave ports in southern Argentina or Chile and cross the Drake Passage to the Antarctic Peninsula—the part of the continent that looks like an arm reaching out to grab South America. A few tourists fly across the Drake Passage and a few go on ships that visit other parts of the continent, such as the Ross Sea, which is almost due south of New Zealand.

The 2,500-mile trip across the empty Southern Ocean to the big U.S. McMurdo Station on Ross Island, brings home the remoteness of Antarctica. The trip is about as long as a flight from Portland, Oregon, to Jacksonville, Florida, mostly over an ocean where a ship is a rare sight, and then over the ice-covered land of Antarctica. At McMurdo, the nearest tree is back in New Zealand. But the place is far from lifeless.

It's on McMurdo Sound; an arm of the Ross Sea, and the sea supports Antarctica's rich, coastal life. During short hikes around McMurdo you are likely to see penguins, skuas (a seabird that looks somewhat like a gull), seals, and even a killer whale from time to time as it pops up from the sound.

CAUTION

Crevasse Caution

Penguins don't live at the South Pole. Polar bears live in the Arctic and penguins live around Southern Hemisphere coasts, including the coast of Antarctica. The nearest part of this coast is more than 800 miles from the Pole, which is almost in the middle of the Antarctic continent. A penguin could reach the South Pole only if someone took it there.

Map of the Antarctic.

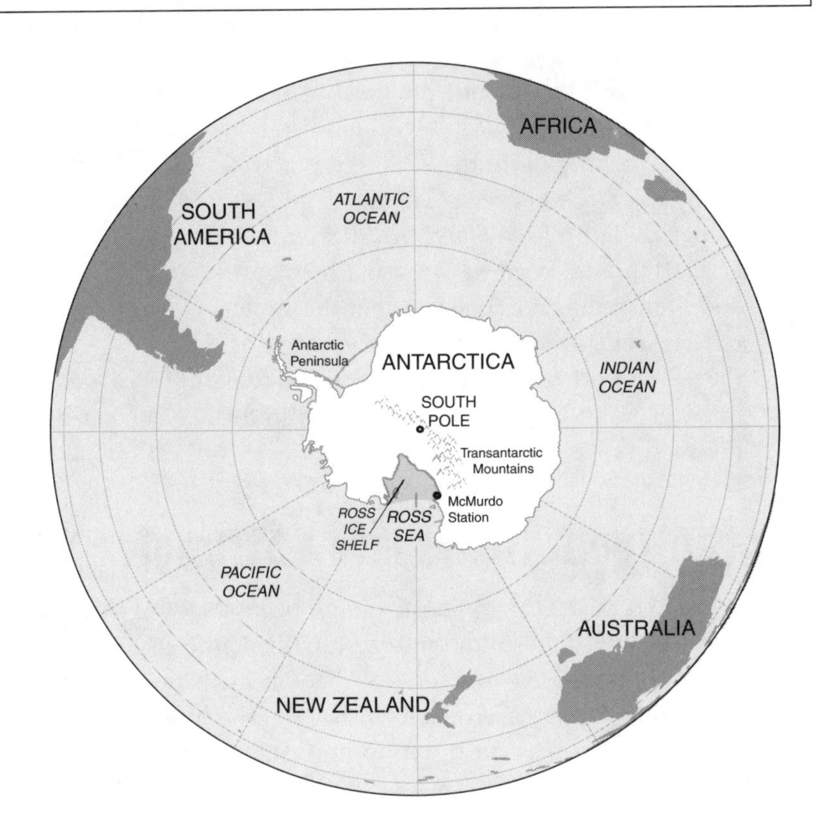

End of the Earth Is Otherworldly

Americans who work in Antarctica call the continent "The Ice." The name is fitting because ice covers around 98 percent of Antarctica, which is roughly the size of the contiguous 48 U.S. states plus about half of Mexico. Once you are more than a few miles inland from Antarctica's coast, the only life you're likely to see other than humans is a wandering skua from time to time. Antarctica is, in effect, a huge ice cube that offers no place for life as we usually think of it to get a foothold. Researchers, however, have found microbes in the ice. Still, the few people who travel to the South Pole see no signs of life other than humans, who would soon die if the technology of keeping warm were to break down.

Ice Chips

December South Pole weather:

- December average high: −15°F (−26°C)
- December average low: −21°F (−29°C)
- December warmest: 8°F (−13°C) (highest ever recorded during the year)
- December coldest: −38°F (−39C)

Frostbite and Sunburn Are Polar Dangers

A typical midsummer day at the U.S. Amundsen-Scott South Pole Station would find temperature hovering around –20°F (–29°C). As one of the ski-equipped LC-130 Hercules transport planes, which almost everyone who goes there takes to the South Pole, slides to a stop on the snow taxiway and one of the crew lowers the ramp at the back of the airplane, the first impression is blinding light. As visitors walk down the ramp onto the packed snow, the next impression of the Pole is numbing cold seeping through the layers of clothing they are wearing.

Before going to the Pole, newcomers are warned to watch out for frostbite, sunburn, and snow blindness. The frigid temperatures will quickly freeze exposed ears and fingers. With a bright blue sky and nothing but thousands of square miles of clean snow surrounding the station reflecting light, anyone who doesn't wear sunglasses or tinted goggles risks snow blindness. (Often called a sunburn of the corona, it's not permanent.) With the sun up 24 hours a day, circling the sky about 20 degrees above the horizon in December, South Pole residents need sunglasses as much at midnight as at noon. They should also apply sun block to any exposed skin. The ultimate natural insult would be to have the tip of your nose both sunburned and frostbitten.

There's More Than One South Pole

Soon after arriving at the Pole, most visitors pose for their "hero" picture at the "South Pole." The place that's often selected first is near where arriving airplanes stop. Poles with flags of the 26 nations that originally signed the Antarctic Treaty surround a foot-wide, reflecting globe on a red and white, waist-high pole. This is the "ceremonial" South Pole, which is close to the station's main entrance, making it easy to dash out for a photo and then back inside.

The geographic South Pole, around 100 feet from the ceremonial pole, is where all of the lines of *longitude* come together at exactly 90 degrees south latitude. The Poles are different because the ice that covers Antarctica at the Pole is slowly sliding toward South America, moving about 30 feet a year, carrying the station with it.

Polar Talk

Longitude is the imaginary lines running from the North Pole to the South Pole, crossing latitude lines at 90° angles. The zero longitude line runs through Greenwich, England. From there, west across the United States to the 180-degree line in the Pacific Ocean, the lines are west longitude. From Greenwich east, across most of Europe and all of Asia to the 180° line, the lines are east longitude.

Each New Year's Day, surveyors from the U.S. Geological Survey, using global positioning system instruments, locate the geographic South Pole and place a marker, with a suitable quotation, on a metal pole. They also put up a sign identifying the Pole. You can see the markers from the most recent years stretching away behind the sign. While the marker shows where all of the lines of longitude meet, it's not necessarily the southern end of the earth's axis of rotation. Because the earth's axis wobbles a little, it might be as much as 40 feet from the marker at any one time.

Explorations

Today, the Geological Survey marker tells you that you're at the South Pole. One of the scientists at the Pole could probably locate the axis of rotation if you wanted to stand there to see if it makes you dizzy. (It won't; you feel no more sense of motion at the Pole than anywhere else.) When the Norwegian Roald Amundsen and the Englishman Robert Scott and their parties arrived in 1911 and 1912, they used sextants to measure the angle between the top of the sun and the horizon. When it was the same all 24 hours of the day, they were within about 1,000 feet of the Pole. Similar methods were used until 1975 when the Geological Survey begin using Doppler satellite measurements, replaced by GPS in 1991.

Life on The Ice Takes a Holiday

Those who live and work at isolated places develop their own traditions, such as the annual Christmas Day "Race around the World" at the South Pole. The race is three laps around a mile-long course that circles both the ceremonial and geographic South Poles. The winners of the race are those who run, or at least jog. Running three miles at the Pole is an accomplishment. The station is 9,450 feet above mean sea level with about 9,000 feet of this being ice. Newcomers find even a short walk leaves them puffing as though they had climbed several flights of stairs, and one of the most common medical problems is altitude sickness. Maybe the difficulty of running is why some "race" by sitting in chairs and sofas on a huge sled being pulled around the course by a Caterpillar Challenger tractor that's normally used to drag heavy loads and the scraper used to groom the station's snow runway. The name of the tractor, selected in a contest, is "The Drag Queen."

More to the Polar Regions Than Circles

As we've seen during our brief look at the Arctic and Antarctic, the world's polar regions are defined by more than frigid air, ice, and snow. The one thing that makes the polar regions different from the rest of the world is the days without sunsets or sunrises.

Even so, it's not completely wrong to think of the polar regions as places with lots of ice. If you were asked to design a logo for a product with the word *polar* in the name, you'd probably come up with something involving ice, or a stylized polar bear or penguin. You wouldn't use a summer scene of the tundra outside Barrow, which is a carpet of green grasses, dwarf trees, and shrubs.

Climate scientists use the term *cryosphere* to describe the parts of the earth covered by ice for part or all of the year. They are interested in the interactions of the cryosphere and "biosphere" (living things), as well as interactions with the atmosphere and oceans. Much, but not all, of the cryosphere is in the Polar Regions. Other parts are the glaciers in the middle latitudes, such as in southern Alaska, and even atop tropical mountains.

As we continue exploring the Polar Regions, we'll spend most of our time north of the Arctic Circle or south of the Antarctic Circle under the midnight sun or in the darkness of the long polar night. From time to time, however, we'll slip across the circles to explore some places that are polar in every way but location.

Polar Talk

The **cryosphere** is the parts of the Earth covered by ice part or all of the year, whether in the polar regions or elsewhere.

The Least You Need to Know

- Unusual patterns of light and dark, as much as cold, define the polar regions.
- The sun rises and sets only once a year at the North and South poles.
- When it's up, the sun circles a pole at the same height above the horizon all day.
- Despite the long winter cold and dark, life thrives in the Arctic and in the ocean around Antarctica.
- Hardly any life is found on Antarctica.
- People have lived in the Arctic around 4,000 years, but Antarctica has never had a native human population.

Chapter 2

Poles Offer Dangerous Weather Under Glorious Skies

In This Chapter

- Why the polar regions are cold
- The polar regions are Earth's air conditioners
- Frigid water warms the Arctic
- Why Antarctica is a cold desert
- Polar air creates stunning skies

A skewed pattern of daylight with some days when the sun never rises and others when it never sets makes the Arctic and Antarctic unique. The weather is also unique, but not only because it's cold. To take one apparent paradox: About 90 percent of the world's fresh water is locked up in Antarctica's ice, yet Antarctica is the world's largest desert.

In this chapter, we'll learn more than how places covered by ice and snow can be deserts. We'll begin by learning why the polar regions are cold and how this affects the rest of the world. From there, we'll look at polar storms and the fierce winds Antarctica produces without the help of storms. Finally, we'll look up at the polar skies to see patterns of light and color rarely seen elsewhere.

Earth's Tilt Chills the Poles

To see how the earth's tilt chills the poles, imagine watching earth from far out in space as it orbits the sun. Now, imagine the earth is marking a long-lasting trail of smoke like from a sky-writing airplane. In a year we'd see that the earth has followed an oval path, almost a perfect circle, around the sun. If we could fill in the area inside the circle, we would create a flat surface—a plane—that goes through the center of the sun and into the center of the earth all along its path.

Now, let's visualize real poles coming out of the earth at the North and South poles. If we made some measurements, we'd see that our imaginary poles make an angle of 23.5° to the plane of the earth's trip around the sun. And an important point, the North Pole is always pointed toward the same far-away star, Polaris, no matter where the earth is in its orbit.

How the earth's tilt causes polar light and darkness.

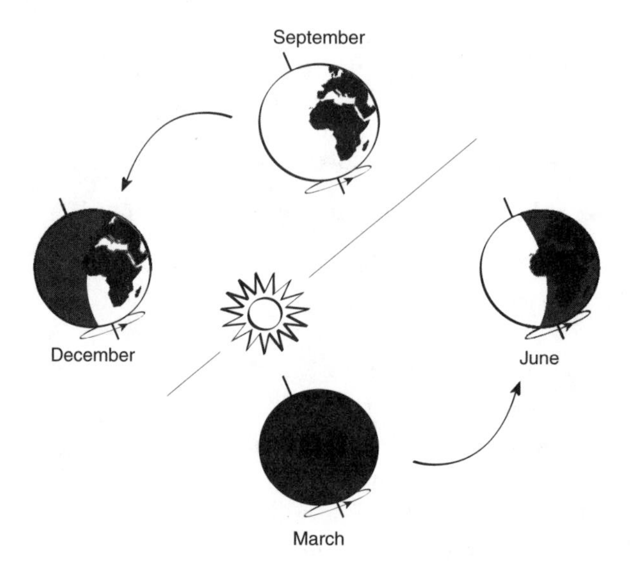

How the Seasons Work

At the end of the year, around December 21, is the winter *solstice* in the Northern Hemisphere. The North Pole is tilted away from the sun while the South Pole is tilted toward it. The southern end of the earth—all the way north to the Antarctic Circle—is bathed in sunlight. At the same time, everything north of the Arctic Circle is dark, no matter what side of the earth it's on. You would also see that every place between the Antarctic and Arctic circles moves into the sun's light and then out of it as the earth spins on its axis.

Around the third week of March, the time of the *equinox*, the earth has moved to a position with its axis pointing neither toward nor away from the sun. The sun is shining directly on both poles. But every place else on Earth, including places not far away from either pole, moves into and out of the sun's light as the earth spins—they are seeing sunrises and sunsets that are close to 12 hours apart everywhere.

Polar Talk

The **solstice** occurs when the sun is above latitude 23.5 degrees north, around June 21, and latitude 23.5 degrees south, around December 21.

The **equinox** occurs when the sun is directly over the equator, around March 20 and September 22.

On June 21, the North Pole is pointing toward the sun and every place north of the Arctic Circle is enjoying 24 hours of sunlight. Now, it's summer in the Northern Hemisphere, winter south of the equator. By September's equinox, the earth is on the opposite side of the sun from where it was in March with day and night nearly equal all over the earth. The Northern Hemisphere is cooling off as the fall days grow shorter and shorter while spring sun is bringing warmth and more sunlight to the Southern Hemisphere.

Summer Sunlight Not as Powerful as Elsewhere

Even when the sun is shining 24 hours a day on the polar regions, less sunlight reaches them than places closer to the equator. An experiment with a flashlight and a dark room shows why. If you point the flashlight down at your feet it makes a small circle of light. You can think of a certain number of photons of light hitting the floor to warm it ever so slightly. Now, shine the light on the floor across the room and it makes an oval of light, with more square inches than the circle at your feet. Yet it has the same number of photons, which means there are fewer to warm each square inch when the light hits at an angle. The spread-out oval is like sunlight hitting the polar regions during summer's 24-hour days. In other words, a lot of the heat from the earth's furnace—the sun—isn't reaching every part of Earth. The polar regions are like a room in a house with a partly blocked heat duct.

Poles Are Earth's Air Conditioners

The earth is always radiating infrared energy toward space. This export of energy keeps the earth from growing hotter and hotter. When the earth's temperature is in balance, the amount of infrared energy the earth sends out into space matches the amount of energy arriving from the sun. You can think of the earth as having a thermostat that keeps incoming and outgoing heat in balance. The debate about global warming, which we look at in Chapter 28, is really an argument about whether humans are turning up the thermostat.

How the Polar Air Conditioners Work

So far we've looked only at Earth's heat arriving as solar energy and leaving as infrared radiation. As anyone who's ever felt an icy blast of polar air on the Great Plains can tell you, there's a lot more to the story. Cold outbreaks and heat waves are all examples of the earth balancing its heat budget by moving air around. Storms are a part of this process. In fact, the contrast between warm and cold air is the major source of energy for *extratropical storms*, the ones that cause *blizzards*. *Tropical cyclones* are also part of the heat-balancing act because they haul warmth toward the polar regions.

Polar Talk

An **extratropical storm** is a large storm that forms over land or a cool ocean. The storm has fronts—boundaries between warm and cold air. Air temperature contrasts are the main energy source.

A **blizzard** is a severe winter storm with low temperatures, winds of 35 mph or faster, blowing snow that reduces visibility to 0.25 mile (400 meters) or less, which lasts at least 3 hours.

A **tropical cyclone** is a large storm that forms over a warm ocean. The entire storm is warm with no fronts. Heat released as water vapor and condensing to form thunderstorms is the main source of energy. Hurricanes and typhoons are tropical cyclones.

Warm air moves into polar regions, loses a great deal of its heat and humidity, and returns to the middle latitudes. This is pretty much what an air conditioner in your home or office does.

The ocean is also involved in the global heat balancing act. The Gulf Stream in the Atlantic Ocean is probably the most famous of the world's ocean currents, but it's only a part of a global system of warm and cold currents that connect the polar regions with the rest of the world. The warm currents, like the warm air that moves into the Arctic and Antarctic, helps make up for the heat the polar regions are losing. The cold ocean currents, many of which run deep under the ocean surface, bring cool water to places far away from the poles.

Frigid Ocean Helps Warm the Arctic

Those who work on the decks of oceanographic research ships in polar waters such as the Arctic Ocean are generally required to wear *Mustang suits*, which are designed to keep the wearer from immediately passing out from the cold if he or she falls into the water. Because salty sea water doesn't freeze until it's around 29°F (–1.8°C), the Arctic Ocean's water is colder than most of us think water ever is. Yet this frigid water actually helps keep large parts of the Arctic warmer than it otherwise would be.

During the winter, a layer of sea ice averaging around 10 feet thick, covers most of the Arctic Ocean. You would think that with months of polar darkness, the water under this ice would become cool enough to freeze; that is, the ice would grow thicker. But as we'll see in Chapter 5, relatively warm water from the Atlantic flows into the Arctic Ocean on either side of Greenland and circles around the ocean under the ice.

Polar Talk

A **Mustang suit** is insulated, protective coveralls, made by the Mustang Survival Corp., and designed to keep someone who falls in the water alive long enough to be rescued.

Although this water is chilly, around 29° or 30°F (1° or 2°C), it represents a lot of heat. Enough of this heat makes it through the ice to keep mid-winter air temperatures above the ocean in the –40°F and C range. While you aren't going to be running around outdoors in your bathing suit, –40° (F and C) isn't bad for a polar region in the middle of winter.

Places near a large body of water, such as the Arctic Ocean, are said to have *marine climates*. Because water is slow to warm up and then slow to give up its heat, marine climates see milder swings between warm and cold both over the course of the year or even a day than places far from an ocean. Arctic winters are cold and stormy, but summers are relatively mild with temperatures often climbing into the 50°F (10°C) range. This extra warmth allows the air to be more humid, which translates into summer clouds and fog.

Polar Talk

Marine climate is the average weather typical of places near an ocean, including relatively small seasonal variations compared with inland places at the same latitude.

The Arctic Is Dry, Even Near the Ocean

In other parts of the world, marine climates tend to be wet because humid air flows off the ocean with moisture for rain for snow. But even ocean breezes in the Arctic are dry because the air is cold, which means it can carry less humidity than warm air. On the average, precipitation—rain or the water from melted snow—in the Arctic ranges from around 8 inches (200 millimeters) to around 16 inches (400 millimeters) a year. Just to give you an idea of how dry this is, Chicago averages 35 inches (900 millimeters) a year.

Go Inland to Cool Off or Warm Up

Colder winter temperatures and warmer summer readings are typical of the *continental climates* of places far from oceans. This is why the coldest parts of the Arctic are far from being the northernmost.

The officially recognized low temperature record for the Northern Hemisphere is –90°F (–67°C) set in both Verkhoyansk and Oimekon in Siberia. Verkhoyansk is only 74 miles north of the Arctic Circle, and Oimekon is 209 miles south of the Circle—it's not even in the Arctic.

Polar Talk

Continental climate is the average weather typical of inland places far from an ocean, including large annual, daily, and day-to-day temperature changes.

Go to Antarctica for an Ice Age

"To visit Antarctica is almost like visiting the last ice age," says Dennis Peacock, of the National Science Foundation's Office of Polar Programs. Not only is it cold, it's almost completely covered by ice. Instead of being primarily an ocean surround by land, like the Arctic, the Antarctic is almost completely filled by a continent.

The ocean that surrounds the continent of Antarctica doesn't take as much chill off the air as you might think. If you turn a globe upside down you see Antarctica and a huge amount of ocean. The other Southern Hemisphere continents are all far enough away to allow the wind and ocean free rein around the world. In addition to kicking up fierce storms and towering waves, the winds blowing around Antarctica shove ocean water from west to east, creating a strong, relatively warm current around the continent, instead of allowing warm water to flow south.

Antarctica Has a Cold, Cold Heart

Even though the ocean around Antarctica is cold, it manages to moderate the climate a little. But warming Antarctica is a struggle since ice covers around 98 percent of the continent. Week after week of 24 hours of sun helps, even with the ice reflecting most of the sunlight that reaches it. Around the edges of the continent, in places such as the U.S. McMurdo Station, temperatures will sometimes approach 50°F (10°C). But even here the average high temperature in January, the warmest month, is only 31°F (1°C).

Ice Chips

The lowest temperature ever officially recorded in the world was –129°F (–89.4°C) at the Russian Vostok Station in Antarctica on July 21, 1983. Vostok is on the Antarctic Plateau at an elevation of 11,220 feet (3,430 meters), which is higher than the South Pole.

Coreless Winter Chills Continent

In most parts of the world the winter has a "core," a period of a few days, in rare cases three or four weeks when the winter is, on the average, coldest. A graph of daily or even monthly average temperatures looks something like stair steps descending to the core and then ascending back to the middle of summer, where you find a brief plateau of warmth. For example, in Minneapols–St. Paul, the core of winter is January 13 through 20 when the normal daily mean temperature—the average of each day's high and low—is 10°F (–12°C).

If we look at monthly average low temperatures for the South Pole station we see it has a coreless winter with the average low falling to –78°F (–16°C) in April and creeping down to an average of –81°F (–63°C) for July, August, and September.

Largest, Windiest Desert Sits on Ice

Places near the Antarctic Coast are among the cloudiest in the world, with around 80 percent of the sky normally being hidden by clouds. As you move inland, however, the skies tend to clear as you ascend onto the high plateau. With all of these clouds, coastal areas receive most of the continent's snow, and even a little rain at times.

Averaged across the continent, enough snow falls each year to melt down to about 5 inches (13 centimeters) of water. But most of this falls near the coast. At places on the Antarctic Plateau, such as the South Pole, a year's snow would melt down to only around 2 inches (5 centimeters) of water. This is why the Plateau is the world's driest, largest, and—of course—coldest desert. On the Plateau, most of the ice that falls is not snow from clouds as elsewhere in the world, but tiny ice crystals that form in clear air and drift down.

Katabatic Winds Are Downhill Racers

Air, like other substances, becomes more dense as it cools. That is, a cubic meter of cold air will be heavier than a cubic meter of warm air. As cold air builds up over the higher parts of Antarctica it begins flowing downhill, creating what are known as *katabatic winds*. As such winds blow downhill, they speed up when they squeeze through valleys or other narrow openings.

Katabatic winds were one of the big, and unpleasant surprises for early twentieth-century explorers such as the Australian, Douglas Mawson, who led a 1911 to 1914 expedition to the part of Antarctica south of Australia. He built a winter camp at Cape Denison in an area that he later called "the Home of the Blizzard." Winds blew continuously faster than 60 mph with gusts faster than 100 mph and to 200 mph at least once during March and April of 1912.

Polar Talk

Katabatic winds are winds created when cold, dense air begins flowing downhill.

You Don't Want to Meet "Herbie"

One of the first things newcomers to the U.S. McMurdo Station are told is to "watch out for *Herbie*." If you look south from the McMurdo area you see White Island and Black Island (two of the most appropriate names in the world, one is black rock, the other ice). Between them is "Herbie Alley." If the view down Herbie Alley is obscured, you had better prepare for a Herbie, which is the local name for a storm that can bring winds up to 100 mph, zero visibility, and life-threatening wind chills. No one today seems to know where the name comes from. Herbies are typical of Antarctica in that they are relatively small-scale storms that can come up quickly endangering anyone who is not prepared. Sometimes such storms, periods of low visibility, or katabatic winds last only a few hours, some times they can continue for days, keeping you from traveling whether by airplane, snowmobile, or even in some cases walking from building to building.

Polar Talk

Herbie is the name for a local windstorm on Ross Island, Antarctica. These storms can bring high winds, zero visibility, and dangerous wind chills to the McMurdo Station.

You Can See Visions in the Cold Air

Usually air temperatures drop as you go higher, but that's not always the case. At times, you can go maybe a few hundred feet up and find that the air is warmer than it was near the ground. This is called an *inversion*. Although inversions occur all over the world, they dominate the winter weather in the polar regions. This occurs as the temperature of the ice or snow plunges, cooling the layer of air near the ground, but not the air a little higher up. At times the temperature might be 35°F (20°C) warmer a couple hundred feet up than on the ground.

Polar Talk

In meteorology, an **inversion** is a departure from the usual decrease of temperature with altitude. In other words, the air is warmer at higher altitudes.

Upside-Down Islands Float in the Sky

One of the most striking affects of inversions is the creation of mirages. As light moves though air of different densities it is bent. An inversion bends the light down so that it tends to follow the curve of the earth, instead of going in straight lines. But when the light reaches our eyes our brain interprets it as having followed a straight line. The result is that we sometimes see things such as islands, ships, or icebergs floating in the air. Inversions cause a variety of effects because the layers of different temperatures aren't likely to be even. They can twist light in several ways. Things can look larger, or taller, or even upside down. Mirages caused by inversions are superior mirages because they make things appear to be higher than they really are.

In addition to creating visions of objects floating in the air, superior mirages can make it possible to see things that are over the horizon; sometimes quite far over the horizon. The history of polar exploration is filled with stories of sailors who saw the images of islands or mountains that were more than 100 miles away. One well-known example is the 1818 spotting of the "Crocker Mountains" in the Canadian Arctic by Sir John Ross and his crew. They were estimated to be about 30 miles (50 kilometers) away. But no such mountains were ever found. One theory is that Ross and his crew saw a mirage of an island around 200 miles (320 kilometers) away.

Ice Chips

The weather station at the McMurdo Station in Antarctica regularly reports "fata morgana all quadrants" (all directions). A fata morgana is a superior mirage with multiple distortions of the images that can create what look like fairy castles in the air or on a distant shore. The name probably refers to the legend of King Arthur's enchanted sister, Morgana.

Sun Dogs Live in Icy Skies

Sunlight shining through ice crystals in the air creates displays of light that can put even the most fanciful fata morgana to shame. Scientists call these displays "halos" because most of them are circles of colored light in the sky, but not all of them. One of the most common "halo" displays is the "sun dog" or "dogs" because they often appear in pairs. They are also known as mock suns or parhelia. They are splotches of colored light on one or both sides of the sun. Sometimes you will see a halo circling through the sun dogs and around the sun. During the winter across most of the United States you can probably see sun dogs once a week or more often. They appear when thin, cirrus, ice-crystal clouds are in the sky. The ice crystals bend sunlight, creating halos, including sun dogs.

While high, thin clouds create halos in most of the world, ice floating only a few feet above the snow in a sky that looks clear is often responsible for polar halos. Such ice crystals help the polar regions produce the world's most elaborate displays of halos, and the South Pole is the best of all places because its ice crystals tend to be pure and homogenous. The Pole's distance from the ocean, which adds sea salt particles to the air, plus the lack of nearby mountains that disturb the air probably account for the Pole's perfect halo-creating crystals.

Warren Tape of the University of Alaska, who studies halos, once said that seeing a really good halo display is "one of life's high points, like getting married or being born."

His comment sums up the feelings of many who venture into the polar regions. The weather can make life dangerous and uncomfortable, but you see things you would never see in places where the sun follows a regular schedule of rising and setting every day of the year, the ice doesn't blind you, and temperatures remain in reasonable ranges.

The Least You Need to Know

- Earth's tilted axis cuts the amount of solar energy reaching the poles.
- The polar regions are the earth's air conditioners, keeping it from growing hotter and hotter.
- The Arctic Ocean keeps the Arctic warmer than it would be otherwise.
- Antarctica is much colder than the Arctic because it's a continent surrounded by an ocean.
- The polar regions are dry, and Antarctica is the world's largest desert.
- Cold polar air creates mirages and stunning halos.

Chapter 3

Frigid Landscapes Aren't All Alike

In This Chapter

- A close look at the tundra
- Where the ground never thaws
- Greenland is different
- Antarctica is more than ice

If you fly over the Arctic in the winter or Antarctica any time of the year, the view from the airplane would be much the same: mostly ice. Even in the summer you see hardly any green in Antarctica. The view of Arctic land is much different in the summer: wet and mostly green. From an airplane it looks almost as wet as the Louisiana Coast where the land merges into the Gulf of Mexico, except you see no trees. In this chapter, we look at the Arctic landscape, the tundra, before moving on to the few parts of Antarctica ice does not cover.

The Tundra Isn't Frozen All Year

The words *frozen* and *tundra* seem to have been linked at birth, and that speaks to a basic fact about tundra. Even in the summer if you dig deep enough on the tundra, maybe just a few feet, maybe several, you'd hit frozen ground—*permafrost*. But in the summer, as anyone who has walked across the tundra in the endless daylight will tell you, it seems anything but frozen. It's green and spongy with puddles of water to avoid.

Polar Talk

Tundra is a treeless area, usually bounded by the ocean, ice-covered areas, or the tree line in polar regions, having permanently frozen subsoil and supporting low-growing vegetation such as lichens, mosses, and stunted shrubs. The tundra is also known, aptly, as the Barren Grounds. Alpine tundra is found high on mountains outside the polar regions.

Permafrost is permanently frozen subsoil that maintains a temperature below 32°F (0°C) continuously for 2 years or longer.

The English word *tundra* comes from a Finnish word meaning "treeless plain," which is a good description. In addition to the vast stretches of mostly flat tundra around the Arctic, "alpine tundra" is found in mountains at all latitudes. These areas obviously aren't "plains," but they are otherwise like Arctic tundra with only small bushes because they have short growing seasons. Unlike deserts, which are too dry to support much vegetation, areas of tundra are too cold for too much of the year for plants to grow very large.

Tundra Blends Into the Forest

Arctic tundra stretches across approximately 5 million square miles (13 million square kilometers) in northern Alaska, Canada, Europe, and Asia. The tundra's northern boundary is the Arctic Ocean, but you can't draw a single latitude line that marks where it ends to the south. In the European Arctic, which is warmed by Atlantic Ocean currents, you can travel to around 71 degrees north latitude before moving onto the tundra. In some parts of eastern Canada, the tundra extends as far south as around 55 degrees north latitude.

As you move far enough south on the tundra, you eventually begin seeing a few, stunted trees and thickets of shrubs, known as *taiga*, the conifer forests that cover large areas of Alaska, northern Canada, Europe, and Russia. The beginning of the forest is the natural boundary of the Arctic, not the imaginary Arctic Circle.

Polar Talk

Taiga (pronounced *ti-ga*) is the Russian word for the thin, northern evergreen forest.

Permafrost Keeps the Tundra Wet

If you considered only annual precipitation, you would expect the tundra to be a desert. For instance, the annual precipitation of 4.7 inches (120 millimeters) in Barrow, Alaska, is less than half the 11.6 inches (297 millimeters) in Tucson, Arizona. Yet the summer tundra around Barrow is a solid green—except for the ponds. You see little green in the desert around Tucson.

Those tundra ponds make the difference. Thank the layer of permafrost a few feet below the surface of the tundra for keeping it from being a desert. When winter's thin covering of snow melts, permafrost keeps the water from soaking into the ground. The resulting bogs and ponds supply plenty of moisture for the grasses, dwarf shrubs, and flowers of the summer tundra.

The Ice Goes Deep

Although the name "permafrost" has a good, solid ring to it, permafrost is anything but stable, as people around the Arctic discovered long ago. If heated buildings are built directly on the ground their heat will begin melting the permafrost under the buildings, causing destructive, uneven sinking.

Explorations

The Alaska pipeline, which runs 800 miles from the oil fields on the Arctic Ocean to the port of Valdez, is a minor tourist attraction because more than half of it is above ground. Oil is pumped through the pipeline at about 120°F (49°C), which keeps it flowing in cold weather. If the pipeline were buried like other pipelines, it would melt the permafrost, allowing the pipeline to sag and break. This is why it's on posts, which are built to be radiators that expel heat to the air instead of flowing into the permafrost.

The Alaska Pipeline, north of Fairbanks, on posts that carry heat away from the ground.

(Darlene Shields)

By definition, "permafrost" describes any soil or rock that remains below 32°F (0°C) for two years or longer. In parts of northern Siberia, permafrost extends almost a mile (1,600 meters) deep. The area covered by permafrost stretches far south from the tundra and is estimated to underlay around 24 percent of the Northern Hemisphere's land, including much of Alaska and maybe half of Russia and Canada. That is, permafrost underlies large areas of northern forests, but soil that thaws in the summer is deep enough to allow the roots needed to support large trees.

Spring Brings Violent Floods

One of the huge differences between the Arctic and Antarctic is that Antarctica has no real rivers that run to the sea; the continent is too frozen. Arctic rivers, in contrast, deliver an estimated 1,008 cubic miles (4,200 cubic kilometers) of water to the Arctic Ocean each year—an estimated 10 percent of the world's river flow. During the winter, many of these rivers freeze solid. Spring floods on Arctic rivers tend to be more violent than elsewhere because with permafrost blocking water from soaking into the ground, little ground water feeds the rivers during winter. The Arctic spring brings sudden surges of water into almost-dry riverbeds. These roaring floods erode riverbanks, increasing the sediment carried to the ocean.

The Ground Is Busier Than You Think

Most of us think of soil as just sitting there while plants grow in it. But the soil of the tundra is anything but passive, and this helps explain the patterns of ridges and cracks that we see as we fly over the tundra. These are called *tundra polygons* or *ice wedge* polygons because of their regular shapes.

> **Polar Talk**
>
> A **tundra polygon** is a pattern on the surface of frozen ground formed above ice wedges. They can have low or high centers. An **ice wedge** is a buildup of ice in frozen soil that is wedge-shape in cross-section.

Ice Wedges Break Up the Ground

The thawing and freezing of the soil opens up cracks filled by water from melting snow. Water expanding as it freezes in the fall pries open the cracks. Cycles of thawing and freezing breaks the ground, creating intersecting cracks, which form the more or less regular polygons. Water fills some polygons, forming ponds and larger lakes, called *thermokarst lakes.* "Karst" refers to a landscape of limestone, characterized by caves, fissures, and underground streams. The lakes on the tundra look like karst, but they are really formed by water's thermal action.

> **Polar Talk**
>
> A **thermokarst lake** is an Arctic lake formed when water holds heat that thaws the permafrost below.

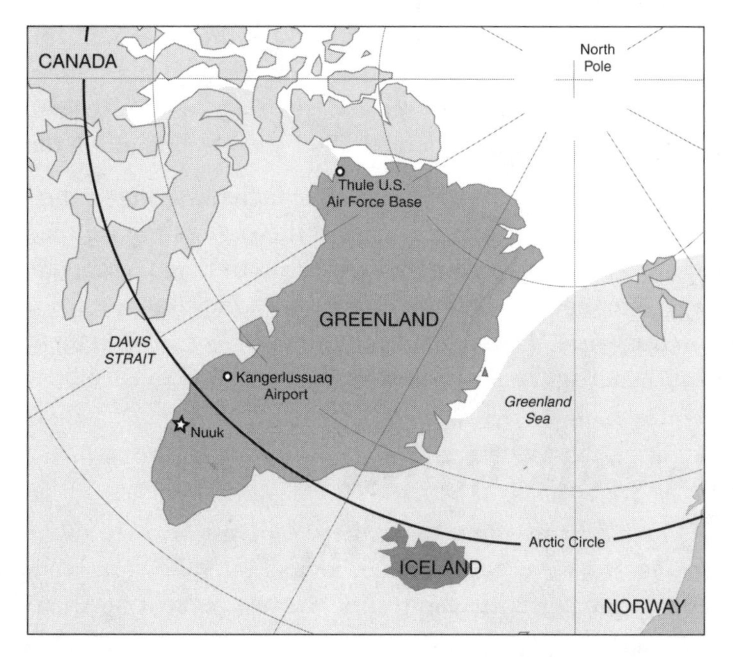

Greenland. Its ice sheet covers all but the very edges of the island.

Greenland Is Like No Place Else

You could argue that Greenland is the most "polar" of all places in the Arctic. All of the Arctic, including Greenland, has cold weather and tundra, but only Greenland has an *ice sheet.* Today, Greenland is the only place in the Northern Hemisphere covered by a huge dome of ice, as large parts of the Arctic and even areas south of the Arctic were in the ice ages that ended around 10,000 years ago. With an area of 840,000 square miles (2.2 million square kilometers), Greenland is the world's largest island, and it's about 85 percent covered by ice.

Polar Talk

The terms *ice sheet* and *ice cap* are often used interchangeably. In this book we'll use the definitions of the National Snow and Ice Data Center at the University of Colorado in Boulder:

An **ice sheet** is a dome-shape mass of glacier ice that covers surrounding terrain and is greater than 19,300 square miles (50,000 square kilometers), such as the Greenland and Antarctic ice sheets.

An **ice cap** is a dome-shape mass of glacier ice that spreads out in all directions. An ice cap is usually less than 19,300 square miles (50,000 square kilometers).

At its longest, Greenland's ice sheet stretches about 1,500 miles (2,500 kilometers) from north to south, and it's about 600 miles (1,000 kilometers) at the its widest. Imagine traveling from Chicago to Phoenix and seeing nothing but ice stretching as far as you can see in all directions, and you'll have some sense of the vastness of Greenland's ice sheet. At the highest part, the ice sheet is around 10,000 feet (3 kilometers) thick.

The 15 percent of Greenland that's not buried under ice includes some of the highest mountains in the Arctic. Rivers, fed by melting around the edge of the ice sheet, flow to the sea, many cutting through long fjords. You won't find trees in Greenland, except in low places that collect water and are protected from the wind. But you will find the oldest rocks anywhere in the world. These date back around 3.7 or 3.8 billion years ago, as close to the earth's estimated age of 4.6 billion years as any rocks yet to be found.

There's More to Antarctica Than Ice

Because Antarctica is "The Ice," you might think there isn't much to say about the continent's land. As anyone who's ever visited Antarctica's Dry Valleys, or seen Mount Erebus—a volcano—even from a distance can testify, the two percent of Antarctica that's not covered by ice has its own fascination. In addition, if you could see through the ice that covers the continent, you would discover a quite different place at the bottom of the earth.

Let's Take a Quick Look at Antarctica

The best way to begin to see what Antarctica is like is to take a quick look at the big picture and then move on to some details (see the following figure). Antarctica covers about 4.5 million square miles (14 million square kilometers). The continent is usually divided into three parts:

- **East, or Greater, Antarctica** includes everything east of the Transantarctic Mountains and is largest part by far. The South Pole and the McMurdo Station are here. Ice, about 1.2 miles (2 kilometers) deep, covers most of East Antarctica.
- **West, or Lesser, Antarctica** is all of the continent, except the Peninsula, that's west of the Transantarctic Mountains. An ice sheet that isn't more than about 6,500 feet (2,000 meters) thick covers most of the region.
- **The Antarctic Peninsula** is the part of Antarctica stretching out toward South America, and is the only major part of the continent extending north of the Antarctic Circle.

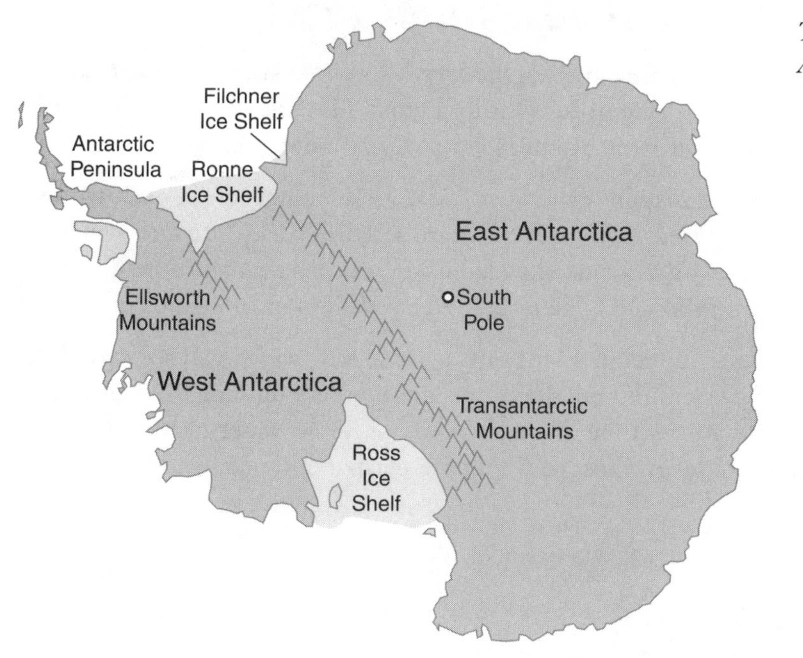

The main features of Antarctica.

Imagine Antarctica Without Ice

If we could remove the ice covering almost all of Antarctica, we'd see that East Antarctica looks like a continent; it's a solid hunk of land with hills and valleys. West Antarctica is a collection of islands with ice filling in the space between them.

Ice Buries Most Mountains

The Transantarctic range is one of the world's longest mountain chains, stretching 3,000 miles (4,800 kilometers) from the Ross Sea to the Weddell Sea. Although snow and ice cover most of the mountains, the few peaks that stick above the ice—called *nunataks*—allow scientists to study the rocks and fossils that help tell the story of the continent's distant past. The highest mountain in the Transantarctic range is Mt. Kirkpatrick with an elevation of 14,850 feet (4,528 meters). Antarctica's highest mountain is the Vinson Massif, which towers 17,840 feet (5,440 meters) above the ice in the Ellsworth Mountains.

Polar Talk

Nunataks are the tops of Antarctic mountains that stick up out of the ice sheet covering most of the continent.

Valleys Have Been Dry Thousands of Years

The largest of Antarctica's ice free areas is the Dry Valleys—a single area with more than one valley—on the west side of McMurdo Sound. The area is about 3 to 6 miles (5 to 10 kilometers) wide and 10 to 50 miles (15 to 80 kilometers) long.

The Transantarctic Mountains block the East Antarctic Ice Sheet from pushing on to McMurdo Sound, helping to create an area of bare soil and rocks: the Dry Valleys. Most of the time, extremely dry winds blow from the Ice Sheet down the valleys. Scientists estimate that no snow has fallen on the Dry Valleys in thousands of years.

Fingers of the East Antarctic Ice Sheet manage to nose part way down some of the valleys. During a few weeks in the middle of the summer, enough ice from these glaciers melts to create, for a brief time, the only rivers of liquid water in Antarctica. These tiny rivers run into lakes, most of which are permanently covered by several feet of ice.

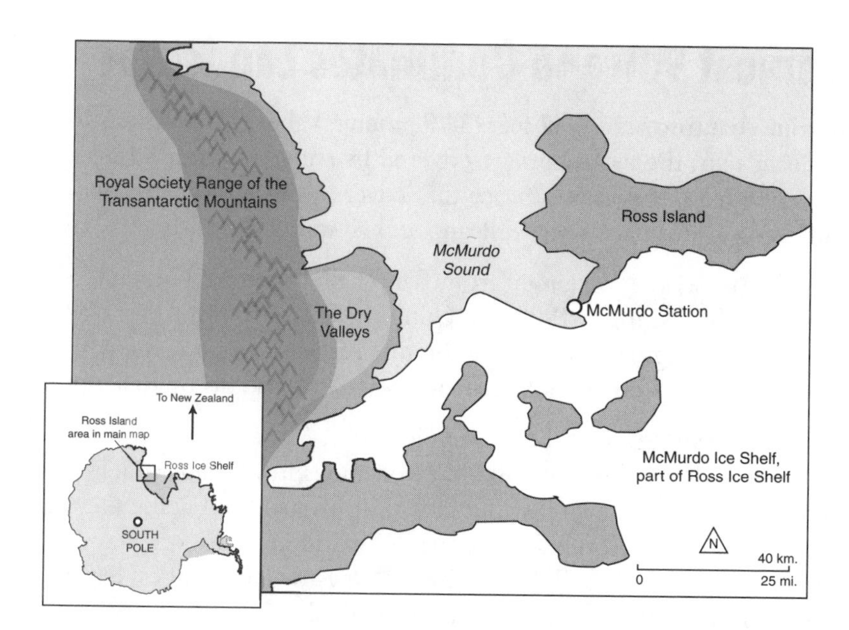

The Dry Valleys.

Salty Lakes, Eroded Rocks, Mummified Seals

A visit to the Dry Valleys gives you a chance to see things found nowhere else on Earth. Some of the valleys' lakes are around 100 feet (30 meters) deep with several feet of ice on top. The lakes are as much as 10 times as salty as sea water and the sun shining through the ice in the summer heats some of them to around 75°F (25°C) while the air temperature is hovering around freezing.

In many parts of the Valleys, you'll find stunning rock fragments called *ventifacts*, which are stones shaped and smoothed by the winds. In most of the world, water erodes rocks, but in arid areas the wind-blown sand and ice crystals ever so slowly grind down and polish rocks. Scientists use ventifacts to figure out the history of glaciers because they can tell the expert eye how long the rocks have been exposed to winds; that is, how long since they were uncovered by a glacier.

If you go to the lower part of the Wright Valley, below Lake Vanda, you will see mummified, rock-hard carcasses of several Weddell seals scattered over several miles. Scientists have found it difficult to come up with good radiocarbon dates for these, but they estimate that the seals are between 2,500 to 3,500 years old. No one knows how the seals, which live in the ocean, traveled several miles inland.

Polar Talk

Ventifacts are rocks in arid areas that have been eroded and polished by wind-blown sand and ice crystals.

World's Southernmost Volcano Dominates Landscape

Mt. Erebus is a white cone that towers 12,447 feet (3,794 meters) above Antarctica's Ross Island. Even though the sides of the mountain are covered by snow, it's as different as anything you could imagine from the snow and ice that covers almost all of the rest of Antarctica. It's the world's southernmost active volcano and Antarctica's most active one.

The first people to climb Mt. Erebus were men from Ernest Shackleton's Nimrod Expedition (the *Nimrod* was their ship) in 1908. Reaching the crater at the top took them five days. Today scientists regularly go to the top of Erebus to work at the Mt. Erebus Volcano Observatory. They don't walk; instead they come and go by helicopter.

Fortunately for the big McMurdo Station about 20 miles away, Mt. Erebus is a well-behaved volcano. In some important ways it is like the volcanoes in Hawaii, which bubble and regularly boil over to send lava flowing downhill toward the Pacific Ocean, but they don't blow their tops off as Mount St. Helens in Washington did on May 18, 1980. Scientists at the observatory can easily track Erebus' day-to-day changes, learning more about how it works than they could at an explosive volcano.

The Polar Regions Are Anything but Boring

One objection that many people have to traveling to the polar regions is that, "There's nothing to see but ice." As we've seen, the Arctic offers a landscape that changes with the seasons as the tundra's snow melts and a sea of green emerges. It's true that Antarctica is mostly ice, but even here the continent offers other vistas. In fact, images of the Dry Valleys and Mt. Erebus remain in the mind of anyone who has ever seen them.

The Least You Need to Know

- Arctic land turns green in the summer while Antarctica's remains mostly ice covered.
- The tundra is a treeless plain with thin soil over permanently frozen ground—permafrost.
- Ice helps creates cracks, hills, and other features on the tundra.
- While ice hides the differences, East Antarctica is like a continent, West Antarctica is a group of islands.
- No rain or snow has fallen on Antarctica's Dry Valleys in thousands of years.
- Mt. Erebus is the world's southernmost active volcano.

Chapter 4

The Strange, Cold World of Ice

In This Chapter

- Water and ice are strange
- How glaciers are created
- Rivers of ice drain ice sheets
- A piece of ice as big as Texas
- It ends with icebergs

In this chapter, we take a brief look at the strange nature of water, which makes ice a very unusual material. We'll see how snow that falls in cold places, such as Greenland and Antarctica, becomes glaciers and ice sheets, which are anything but passive "ice cubes" that just sit in place. We'll see how the ice builds up and moves to the sea. In Antarctica, the ice forms floating "ice shelves," including two as big as Texas. These ice shelves along with glaciers in Greenland are the sources of icebergs that melt, returning the water to the oceans that held it before it fell as snow to make the ice.

Water Is Common and Complicated

Water is so common that unless you are a chemist, you probably don't realize how strange it is compared to the other substances that make up our world. The properties of water that make it so different, and so important to life, would make a book in themselves. We aren't going to attempt to do that. Instead, we'll take a look at what happens as water freezes into ice. Most people know, without having studied chemistry, that water is H_20. You probably also know that this means a water molecule is made of two atoms of hydrogen and an atom of oxygen. What chemists call a covalent bond holds the two hydrogen atoms to the one atom of oxygen. "Covalent" means the atoms are bonded by sharing electrons.

In water molecules, the sharing isn't equal. The oxygen atom attracts electrons more strongly than the hydrogen atoms, making the oxygen atom slightly negative (electrons have negative charge). Hydrogen atoms are slightly positive. Thus, water is a "polar molecule" with negative and positive sides.

Water Is Familiar in All Three Stages

Water is very unusual in that it's found in the three states of matter—solid, liquid, and vapor—at Earth's "ordinary temperatures." Liquid water is the most familiar; it's the expensive stuff we buy in plastic bottles or drink from the tap. Ice is solid water. You can't see water vapor because it's the invisible, gas form of water. When there's enough water vapor in the air, we feel it as humidity.

Ice Chips

Temperature is a measure of the average speed of the atoms and molecules in whatever is being measured. The higher this average speed, the higher the temperature. If all atomic or molecular motion stops, the temperature would be "absolute zero," which is –459°F (–273°C).

The average speed of the molecules or atoms determines whether a material is gas, liquid, or solid. In a gas of any kind, including water vapor, the molecules are zipping around too fast for molecules to "hook up" for long. As the temperature cools, the molecules slow down enough to be attracted to each other, but are still free to move around. Finally, when it's cold enough, the molecules lock together to form solids.

Water Grows Less Dense with Cooling

For almost all substances, the density increases as the substance cools. That is, the molecules draw closer together with the result that a cubic inch of cold stuff weighs more than a cubic inch of the same stuff when it's warm.

Water is very different, however. As water cools toward 32°F (0°C) it actually becomes less dense after it drops below about 39°F (4°C). The reason is that the hydrogen bonds between the positive and negative ends of the water molecules begin to take over and form a crystal lattice. These bonds keep the water molecules a certain distance apart, which doesn't happen when water molecules are moving around in a liquid.

How to Understand Low-Density Ice

One way to help you understand why water's density decreases as it turns to ice is to imagine a platoon of 48 soldiers standing around outside, some close together, others moving around.

The sergeant shouts: "Fall in!"

The soldiers line in up in their assigned places in four rows of 12 with those at the right ends of the back three rows standing arm's length from the soldier in front.

The sergeant shouts: "Dress right, dress!"

The heads of all the soldiers except the four at the right end of each row snap to the right as each soldier, except those on the left end of each row, flings his left arm up to touch the right shoulder of the soldier next to him. Everyone moves to be at arm's length from the next soldier and the one in front and also lines up with them.

"Ready … front!" the sergeant shouts.

Heads snap forward and arms drop, leaving the 48 soldiers standing in straight rows, each one an arm's length from those around him.

For water, the "dress right, dress" command comes when it cools to 39°F (4°C). Once this happens, the water molecules begin forming in six-sided crystals. The shape of water molecules and the way the hydrogen and oxygen atoms bond creates the six-sided crystals.

Ice Floats, and That's Important

Water's expansion as it cools and freezes is why it breaks jars and water pipes. Obviously the hydrogen bonds are strong. Water's decreasing density as it approaches freezing also makes ice less dense than water. A cubic inch of ice, weighs about 93 percent as much as a cubic inch of water. Because ice is less dense than water, the ice floats. If ice didn't float, it would form on the bottoms of rivers, ponds and lakes. Marine life would be cut off from food on the bottom and ice skating would be impossible.

An important consequence is that when floating ice melts, it will not cause sea level to rise. When ice is floating it is already displacing, or pushing up, an amount of water that's almost equal to the amount of water that will be added when it melts. The ice of Antarctica's large ice shelves caused sea levels to rise when they first pushed out to float. When they melt, they won't cause further sea-level rise. When an iceberg falls into the water, it causes the sea level to rise then, not when it melts.

Falling Snow Builds Glaciers

In most of the world's snowy places, spring and summer warmth melts the snow. But in large parts of the polar regions and many high mountains where the annual average temperature is colder than 32°F (0°C), at least some of the snow does not melt. The next year more snow falls and fails to melt, snow piles up over the years.

As snow piles up, a lot of things begin happening under the fresh snow on top of the pile. First, the snow on top is pressing down on the old snow, which compresses it. It's easy to imagine the weight of the snow above breaking the points of snow crystals. At the same time, water vapor evaporates from crystal points and freezes on the smoother parts. These actions create spherical crystals that are more or less the same size, about as big as grains of rock salt. After a year, this ice with the rounded crystals is called *firn*. At this stage, spaces between the crystals allow air to move in and out of the snow and firn.

Polar Talk

Firn is rounded, small grains of snow that are bonded together and which are at least a year old.

Pressure Creates Glacier Ice

The weight of more and more snow on top squeezes out more and more of the air between the crystals in the firn. The air is never completely squeezed out, but ends up trapped as bubbles. The pressure causes ice crystals to grow larger and larger, in some cases becoming more than a hundred yards long. By now, the ice is considered to be *glacier ice* or *glacial ice*. How long this takes depends on the temperatures and amounts of falling snow. At a cold place with very little snow, such as the South Pole, the change from firn to glacier ice can take more than 1,000 years. Fresh fallen snow can have a density of less than 5 pounds per cubic foot (80 kilograms per cubic meter). Glacial ice has a density of at least 55 pounds per cubic foot (800 kilograms per

Polar Talk

A **glacier** is a body of natural ice on land that flows or has moved in the past. **Glacier ice (glacial ice)** is well-bonded ice crystals compacted from snow.

cubic meter). But even at the bottom of the deepest ice sheet, the ice's density is never more than that of fresh water, which is 62 pounds per cubic foot (1,000 kilograms per cubic meter).

Explorations

Icebergs, glaciers, and other large pieces of ice are often marked by streaks or even large areas of deep blue. Snow looks white because bubbles of air between ice crystals scatter, or reflect, all of the wavelengths of sunlight back to our eyes, which we see as white. When ice is squeezed deep in a glacier, the remaining, tiny air bubbles scatter little light, allowing sunlight to penetrate deeper into the ice. Ice crystals absorb six times as much light at the red end of the spectrum than at the blue end. With the ice absorbing most of the red light, only blue is left to reflect back to us.

Where to Find Glaciers

Glacial ice covers about 6 million square miles (16 million square kilometers) of the earth, or about 10 percent of the earth's land area. Most of this ice—about 5 million square miles (13.5 million square kilometers)—is in Antarctica. About 700,000 square miles (2 million square kilometers) covers most of Greenland while the remaining 200,000 square miles (500,000 square kilometers) is spread around the world, mostly in glaciers in the middle latitudes, especially the cooler parts such as Alaska, the Canadian Rockies, and the higher mountains of Asia, Europe, and South America.

Glaciers Take Their Ice Somewhere

When enough snow and ice pile up, the ice begins acting like a liquid; a very thick liquid. Since the pressure is causing the movement, the ice on the bottom is moving. This begins happening when the ice is about 150 feet (50 meters) thick. Many glaciers, especially in mountain valleys, flow downhill, but that's not necessary. When the ice is thick enough, the glacier can even move uphill.

To be considered a glacier, the ice has to be moving, or at least have moved in the past. In other words, a glacier is nature's way of collecting snow in some places and carrying the water the snow contains, somewhere else. Eventually, the ice melts or breaks off because glaciers don't go on forever. How fast a glacier's ice moves depends on many factors, such as whether the bottom of the glacier is frozen to the rocks, or is slipping across them, maybe with the help of wet sediments under the glacier.

Ice Reaches the End of the Road

Most of the ice that's locked up in glaciers ends up melting with only a small amount evaporating directly into water vapor. A glacier's ice can melt on land to send water flowing down streams and rivers to the ocean. If a glacier reaches the ocean, a bay, or a fjord, icebergs break off and drift away. Sometimes, especially around Antarctica, ice forms ice shelves at the ocean's edge. These are big shelflike pieces of ice attached to the ice sheet at the land end that stick out to float on the ocean.

Where Greenland's icebergs go.

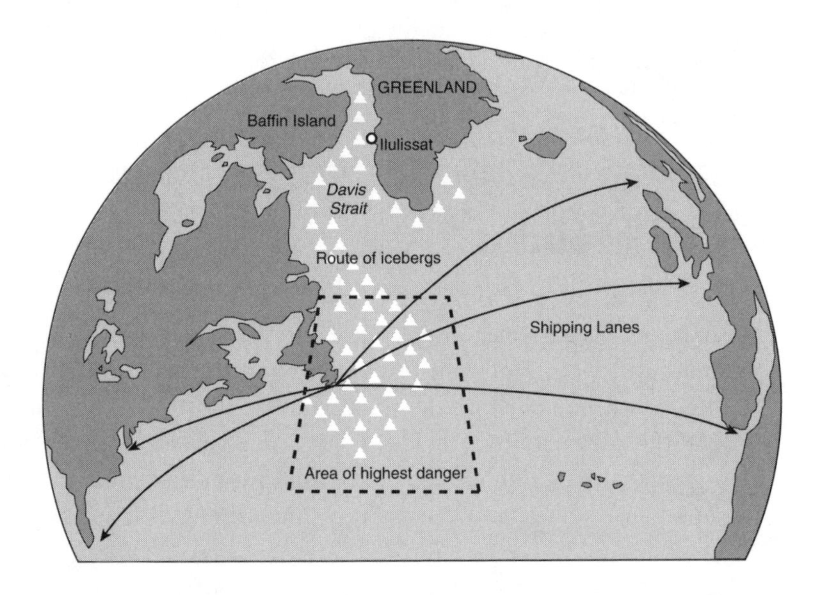

Atlantic icebergs are the most dangerous because they float into the paths of ships. Almost all these come from Greenland's ice sheet. The sinking of the *RMS Titanic* on April 15, 1912, with the loss of 1,517 people illustrates not only the dangers of icebergs, but also how the polar regions affect other parts of the earth. The Arctic Ocean exports cold water into the Atlantic as the Labrador Current, which flows south to meet the warm Gulf Stream. A temperature difference in the water and the air right above it, which reaches 35°F (20°C) at times helps create dense fog and strong storms. The Labrador Current also hauls icebergs from Greenland south to the Grand Banks off Newfoundland, about 1,000 miles (1,600 kilometers) south of the Arctic Circle. Ships following the shortest routes between Europe and North America cross this area.

Ice Chips

Antarctic icebergs can be the size of a small U.S. state. Arctic bergs are much smaller, closer to the size of a medium-size office building. The U.S. Coast Guard says a "medium" berg is from 51 to 150 feet (15 to 45 meters) high, including the part under water, and 201 to 400 feet (61 to 122 meters) in diameter.

After the *Titanic* disaster, the world's major shipping nations signed the first Safety of Life at Sea convention in 1914, which created the International Ice Patrol. The Patrol continues today with U.S. Coast Guard airplanes making regular flights to spot and track icebergs during the season, which usually runs from February through July. The Patrol has been successful in keeping ships from being sunk by icebergs in the area it covers, but the danger is still very much alive elsewhere. For instance, on January 30, 1959, a Danish ship, the SS *Hans Hedtoft*, hit an iceberg in a snowstorm off Cape Farewell, Greenland, and sank, killing all 95 passengers and crew.

Ice Rules in Antarctica

As we've seen, "The Ice" is an appropriate name for Antarctica since ice covers around 98 percent of the continent. As you fly over Antarctica, the ice looks almost like an endless, snow-covered plain. The East and West Antarctic ice sheets are very different, however, and when we look under the apparently featureless ice of Antarctica, we see that a lot is going on.

East Antarctica Holds Most of World's Ice

The East Antarctica Ice Sheet is the 800-pound gorilla of the world's ice. It consists of 6.2 million cubic miles (26 million cubic kilometers) of ice, which is close to 80 percent of all of the world's ice, including the ice elsewhere in Antarctica. It would cause global sea levels to rise by 215 feet (65 meters) if it all ever melted. Its ice averages around 6,500 feet (2,000 meters) thick and its highest point, called Dome Argus, is 13,222 feet (4,030 meters) above sea level.

Fortunately, all of this ice is what glaciologists call a *land-based ice sheet*. (The Greenland Ice Sheet is also land-based.) This means that the bottom of the ice sheet is above sea level. Land-based ice sheets are more stable than *marine-based* ice because if a land-based ice sheet starts melting, sea water wouldn't be able to flow under it to speed disintegration from the bottom.

Polar Talk

When the bottom of an ice sheet is generally above mean sea level it is called a **land-based ice sheet.** When it is generally below mean sea level it is called a **marine-based** ice sheet.

West Antarctic Sheet Is One to Watch

With approximately 783,000 cubic miles (3.2 million cubic kilometers) of ice, the West Antarctic Ice Sheet has only about 11 percent as much ice as its big brother in East Antarctica. Since the late 1970s, however, scientists have been paying a lot more attention to the West Antarctic Ice Sheet because it's the one, at least in theory, that could melt during this century. If this happened, it would raise global sea levels by an estimated 20 feet (6 meters). This would cover most of the world's low-lying islands and wipe out a lot of waterfront property on all the continents.

Rivers of Ice Flow Between Ice Banks

Antarctica's ice is moving slowly to the ocean, but not as one mass. Instead, it converges in streams of ice that move though the huge mass of ice that's pretty much staying in the same place, at least for now. These ice streams are around 300 miles (500 kilometers) long and from 10 to 60 miles (20 to 100 kilometers) wide. The ice moves around 3 to 6 feet (1 to 2 meters) a day, but streams speed up and slow down. At least one stopped about 150 years ago and another has been slowing down in recent decades.

CAUTION

Crevasse Caution

If you ever have a chance to walk on an ice sheet or glacier, watch out for crevasses, which are cracks in the ice that can extend hundreds of feet down. Sometimes "snow bridges" hide crevasses. You step on what looks like snow and fall through. Fortunately, glaciers or ice sheets crack into crevasses only in certain places. Get expert advice to learn where the danger is.

Some Ice Streams Slide on Water

As amazing as it might sound, the bottom of an Antarctic ice sheet could be much warmer than the air at the top. If you could go down into one of the ice sheets you'd find temperatures range from maybe –5°F (–20°C) to –20°F (–30°C) or colder while the air above might be –40°F and C. Close to the bottom of the ice sheet, however, you'd find the temperature going up, in some cases reaching ice's melting point. This warmth comes from heat seeping up though the rocks under the ice, or the friction of ice moving across the rocks.

Even with warming from below, the bottoms of many ice sheets and glaciers remain well below freezing and the ice at the bottom is frozen to the rocks. These ice sheets or glaciers can still move, but movement occurs as the ice deforms; that is, the ice acts like a very thick liquid.

At other times, such as with the ice streams of West Antarctica, the bottoms of parts of the ice sheets are warm enough for ice to melt. In these areas, the ice can move faster than ice that's frozen to the rocks. A thin layer of water lubricates the bottom of the ice sheet. Sometimes the bottom of the ice sheet sits atop tiny rocks or soil, which makes it even easier for the ice to slide when water is mixed in. Sometimes the water can flow in under a part of the ice sheet from upstream where it has melted. Changes in the amounts of water under different parts of the Ice Sheet account for ice streams that speed up, slow down, stop, and start over the years.

Ice Chips

As the pressure on ice increases, the temperature at which it begins melting goes up. For example, at the bottom of a 1.36 mile (2,200 meter) thick ice sheet, ice begins to melt at 29.12°F (–1.6°C) instead of at 32°F (0°C). The thicker the ice, the colder the melting point.

"States" That Float on the Ocean

When James Clark Ross with his two ships sailed into the Ross Sea (later named after him) in 1841, they saw a wall of ice around 100 feet high that stretches out of sight in both directions. They called it "The Barrier," but we now call it the Ross Ice Shelf, named after: guess who? The ships sailed close to the barrier and measured the depth of the water, which showed the ice was floating. At least some of those aboard, such as Ross, knew that any ice floats with about 90 percent under the water. Thus, they realized that the barrier extended 900 or so feet under water.

Most of the moving ice on the West Antarctic Ice Shelf ends up as part of either the Ross or Ronne ice shelves. Ice from the East Antarctic Ice Sheet and other parts of the continent feeds scores of smaller ice shelves around Antarctica, which make up close to 50 percent of the coast. The large shelves, such as Ross, are maybe 4,000 feet (1,200 meters) thick where they are anchored to the land.

Crevasse Caution

You can't take the rule that 90 percent of any piece of floating ice is under water too literally. Around 90 percent of the ice's mass, not its height, is under water. If you see a 10-foot, pointed iceberg, the underwater part might be only 50 or 60 feet deep, but wider than the part you see.

As ice flows onto the shelves, the shelf "grows" out into the ocean. Ted Scambos, a scientist at the National Snow and Ice Data Center in Boulder, Colorado, estimates that on the average, the Ross Ice Shelf grows approximately 3,000 feet (1,000 meters) a year. Obviously, this can't go on forever, otherwise the Ice Shelf would have nosed past New Zealand long ago.

Bergs Help Keep the Ice in Balance

The ocean end of an ice shelf is sticking out into water that's relatively warm—at least 29°F (–1.6°C)—which would tend to slowly melt the ice even if nothing else were going on. Other things are going on, mainly the normal rise and fall of the ocean's water caused by the tides. Ross Sea tides aren't huge, maybe from a couple inches to around three feet (5 to 100 centimeters). But that's enough to crack ice shelves. Think of bending a metal coat hanger back and forth until it breaks. It takes decades, but eventually these cracks break, allowing a huge chuck of ice to float away as one of Antarctica's unique icebergs.

Explorations

Antarctic icebergs have carried away important artifacts from the continent's history. Between 1928 and the mid 1950s, U.S. expeditions led by Admiral Richard Byrd set up five Little America camps near the eastern end of the Ross Ice Shelf. Sometime in the 1960s, one iceberg, maybe more, calved off from the same area taking what was left of the camps. Those bergs or earlier ones also carried away the camp Roald Amundsen set up for his successful South Pole trip in 1911. Supplies, equipment, aircraft, and vehicles explorers left behind are now somewhere on the bottom of the ocean where the melting icebergs dropped them.

Is the Ice Melting?

Is polar ice melting? This is one of the most common questions that people have about the Arctic and Antarctic. The answer is "yes." The answer would have been "yes" around the beginning of the last Ice Age around 100,000 years ago. A related question is: "Is polar ice growing?" As with the melting question, the answer is "yes" as it probably has been since the time in its distant past when Earth had no permanent ice. Each year, snow falls on places where it doesn't melt. Glaciers, ice caps, and ice sheets carry the resulting ice to places where it melts. The question, which we return to in Chapter 27, should be: "Is the world's ice melting faster than it's growing?"

The Least You Need to Know

- Glaciers and ice sheets build up when snow doesn't melt from year to year.
- As snow piles up, its weight changes the old snow below into glacier ice.
- Glaciers carry ice from where snow accumulates to where ice melts or breaks off as icebergs.
- Land-based ice sheets are more stable than ones with the bottoms below sea level.
- Ice shelves are huge areas of floating ice.
- The question about polar ice should be: "Is more melting than being added by falling snow?"

Chapter 5

Icy Oceans Have Global Reach

In This Chapter

- The freezing and melting of Antarctic ice is a big deal
- A layer of salty water protects Arctic Ocean ice
- Arctic Ocean water helps drive global currents
- The up, down, and around of Antarctic water
- Polar ice has global effects

In Chapter 4, we looked at ice that forms on land. Even a large Antarctic iceberg that floats in the ocean for a decade after breaking off from an ice shelf is still land ice. The term *sea ice* is reserved for ice that forms from ocean water, not ice that forms elsewhere and ends up floating in the ocean. In this chapter, we'll look at not only the sea ice that forms on the ocean around Antarctica and Arctic Ocean ice, but also at the oceans themselves, and their unique currents that are a part of the flow of water that links all of the world's oceans.

How Sea Ice Forms and Grows

If you were on a ship sailing on the Arctic Ocean or the Southern Ocean in late summer, you'd feel the air growing colder each day as the sun rises for a shorter and shorter time. One day you'd look out on what had been a mostly ice-free ocean to see the water has taken on an oily look. Ice is beginning to form.

Salt lowers the freezing point of water. Ocean water consists of about 96.5 percent water and around 3.5 percent dissolved solids that we call "salt." This is enough salt to lower the freezing point of the ocean's water to around 29°F (–1.8°C). The exact freezing point depends on how much salt is in the water. As water cools to the freezing point, ice crystals begin forming, and as enough form, they give the water a soupy, or greasy look.

Ice Chips

The southern parts of the Pacific, Indian, and Atlantic oceans surround Antarctica, and all have much in common, including the current that circles Antarctica. In this book, as in many others, we use "Southern Ocean" to describe the water around Antarctica.

Eventually, as more and more ice crystals form and stick together, sheets of ice called *nilas* form. These are thin enough to bend with the waves without breaking. As chunks of floating ice become thicker, the pieces bump into each other and can begin to look like pancakes with slightly upturned edges. This *pancake ice* grows into *ice floes* that become thicker and larger as the air cools.

The colder the air, the more quickly ice forms. As water freezes, the ice crystals have no room for the salt in the water, and it's expelled. When the air is very cold, however, the ice can form so quickly that salt is caught between ice crystals, which means the ice will have pockets of brine—water with salt mixed in—trapped in it. If the lasts a year or more, the brine normally seeps out. Water from old *sea ice* can be drinkable.

Polar Talk

Nilas is a thin elastic crust of ice, easily bending on waves. **Pancake ice** is mostly circular pieces of ice from one to 10 feet (30 centimeters to 3 meters) in diameter, and up to 4 inches (10 centimeters) thick, with raised rims caused by the pieces hitting one another. Any contiguous piece of sea ice is an **ice floe,** which can be as small as 60 feet (20 meters) across and larger than 6 miles (10 kilometers).

Any form of ice found at sea that originated from the freezing of seawater is called **sea ice.**

Ice Creates the Globe's Biggest Seasonal Changes

By April of each year as the air over the ocean around Antarctica is becoming really frigid and sea ice formation has reached its stride, 30 to 40 square miles (75 to 100 square kilometers) of new sea ice forms each minute. This advance of sea ice, which is arguably the world's biggest seasonal climate event, goes mostly unseen by humans. By late September, as days grow longer than nights over the Southern Ocean, sea ice covers approximately 7 million square miles of water. The ice covering the bottom of the world—including Antarctica's ice sheets—has more than doubled in size.

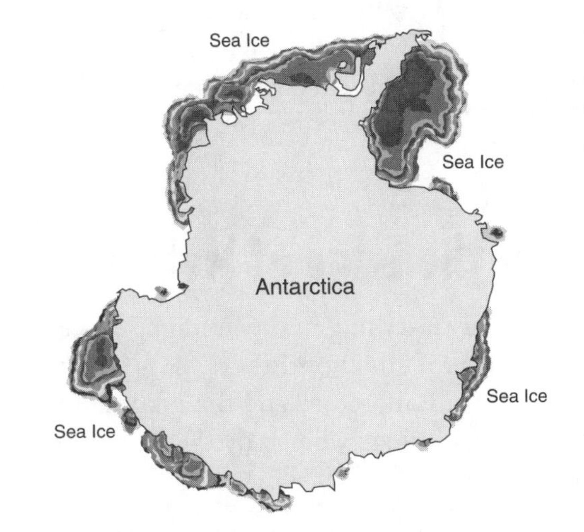

Typical late-February sea ice extent, 1.5 million square miles (4 million square kilometers).

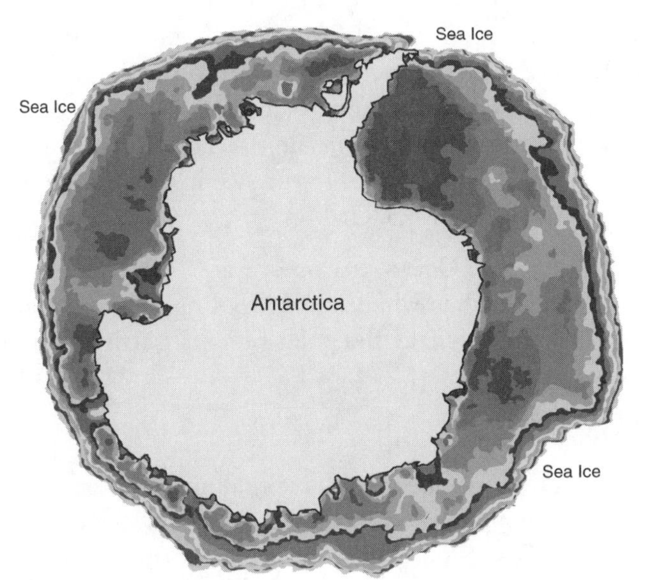

Typical late-September sea ice extent, 7.3 million square miles (19 million square kilometers).

As the sun returns to southern part of the globe, the 4.5 million square miles (14 million square kilometers) of ice covering Antarctica, and the more than 7 million square miles of sea ice around it, reflect away sunlight, delaying spring warming. When melting reaches its stride, enough ice to cover Rhode Island melts each hour. But the solar heat that goes into this melting is energy that's not available to warm the ocean and the air above it.

Polar Talk

In meteorology, **sublimation** is the change of state of water directly from ice to vapor without first melting into a liquid.

Near the continent, the cold, very dry winds blowing down from the ice sheet and over the ice causes some of the top layer of ice to *sublimate* directly into water vapor without first melting into puddles of water. By the end of the Antarctic's short summer in February, approximately 80 percent of the ice that surrounded the continent only 5 months before is gone.

Arctic Ice Floats on the Edge of Melting

Arctic Ocean sea ice isn't the sure thing that you might expect. "There's enough warm water (in the Arctic Ocean) to melt all of the ice," says Robin Muench, an oceanographer with the Naval Research Laboratory and the National Science Foundation. Fortunately, he adds, oceanographers who study the Arctic don't expect this to happen soon, as we'll see in Chapter 27.

Atlantic Ocean water that flows into the Arctic Ocean between Greenland and Norway is coming from the part of the Atlantic that's warmed by the Gulf Stream and its extension, the North Atlantic Drift. Water that flows into the Arctic makes a slow journey around the ocean as a complex system of currents. Vast amounts of water from rivers and a relatively small amount from the Pacific, which comes through the narrow Bering Strait, join the Atlantic water. In the Arctic Ocean, the Atlantic water cools to around 32°F or 33°F (0°C or 0.5°C).

After maybe 25 years in the Arctic Ocean, a drop of water will flow back into the Atlantic, cooler by a few degrees than when it left. Cooling of ocean water during its trip around the Arctic Ocean helps make the polar regions Earth's air conditioners.

Salty Water Preserves the Ice

If you lowered instruments that measure the water's salinity and temperature through Arctic Ocean ice, the instruments would first find a layer of water with little salt, which freezes at around 29°F (–1.5°C). This is where ice forms. Farther down, the water becomes saltier and saltier, but the temperature stays about the same as higher up. Finally, you reach a layer of water that's salty, but warmer, with temperatures slightly above 32°F (0°C). This is mostly Atlantic Ocean water.

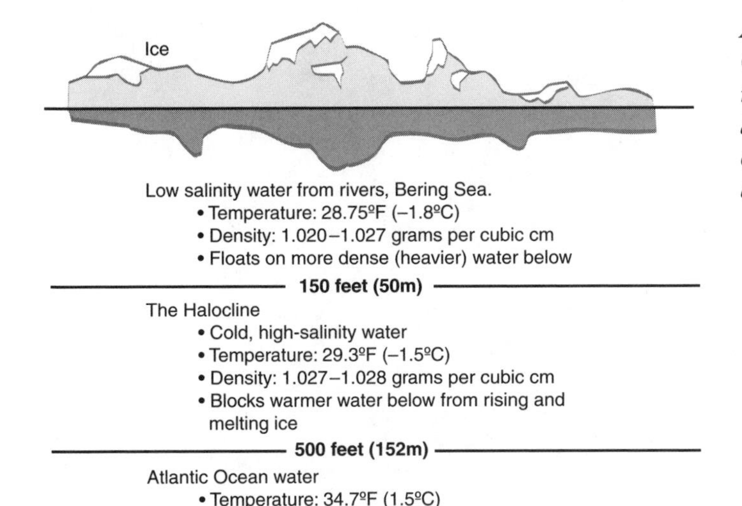

A cross-section of Arctic Ocean water showing typical temperatures and density at different depths. The density of fresh water is 1 gram per cubic centimeter.

The layer of cold, salty water near the top is called the *halocline*. It shields the ice at the top from the warm water below—the water with enough warmth to melt all the Arctic's ice. The less dense—lighter—water on the top floats on the more dense—heavier—halocline water.

Polar Talk

Halocline is a layer of water in which the salinity increases rapidly with depth. *Halo* comes from the Greek word for "salt," and *cline* from a Greek word for "lean." It is generally used in English for a continuum between two extremes.

Fresh water from the major rivers around the Arctic has to be a key player in maintaining the halocline, as it spreads out over the top of the ocean—floating on the halocline—and is the first to freeze in the fall. When Arctic seawater freezes, it rejects salt that sinks to help maintain the halocline.

Arctic Ice Has a Chance to Grow Old

By late February, when those who live around the Arctic are beginning to see the first signs that winter might end some day, sea ice covers almost all the Arctic Ocean, and spills out of the Arctic down both sides of Greenland and along Canada's northeastern coast to northern Nova Scotia.

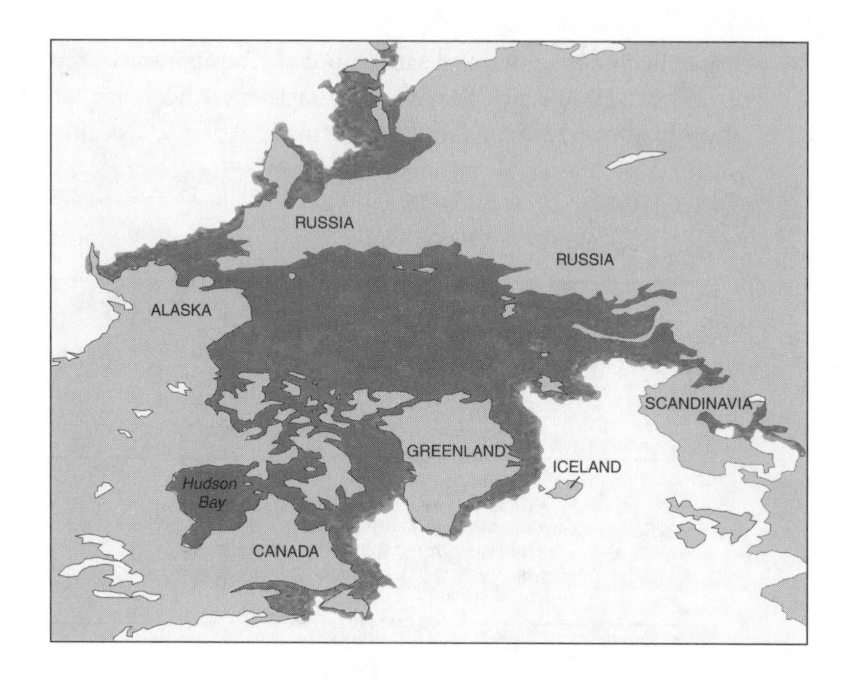

Mid-February: Ice covers approximately 5.8 million square miles (15 million square kilometers) of ocean.

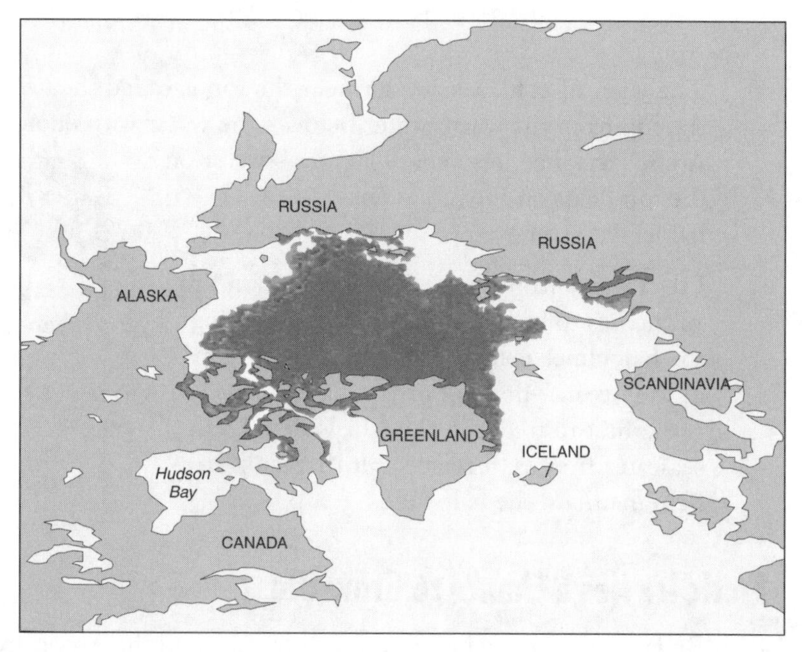

Mid-September: Ice covers approximately 3.5 million square miles (9 million square kilometers) of ocean.

Unlike around Antarctica where storms are free to push the ice away, Arctic Ocean ice leaves the ocean mostly between Greenland and Svalbard, meaning it isn't spread out as much as Antarctic sea ice to be melted by warming ocean water. Only about 40 percent of the Arctic's sea ice melts each summer. Because so much less Arctic Ocean ice melts each year, you see more multi-year ice in the Arctic than around Antarctica. In fact, around half of the Arctic's ice is multi-year ice, which can grow to be about 10 feet (3 meters) thick.

Currents and winds slowly push the ice, but this doesn't mean it can't exert great force when floes run into one another. Colliding ice floes push up *sea ice ridges* and, at the same time, shove ice underwater to form *sea ice keels*, which are usually 30 to 75 feet (10 to 25 meters) underwater.

Polar Talk

A **sea ice ridge** is a line or wall of broken ice forced up by pressure. The submerged, broken ice under a ridge, forced downward by pressure is a **sea ice keel.**

Arctic Reaches Out to the World's Oceans

Water that has spent maybe a quarter of a century circling the Arctic Ocean doesn't just meekly merge with the water of the Atlantic when it emerges from the Arctic. Its trip around the Arctic Ocean has cooled the water, but melting sea ice has made it fresher. Once in the Atlantic, this water can mix with saltier water that the Gulf Stream and North Atlantic Drift have carried north. The resulting cold, salty water is dense and therefore begins sinking toward the bottom of the ocean, forming what oceanographers call *North Atlantic Deep Water.*

Oceanographers now believe that water sinking in the ocean south of Greenland is an important driver of a global system of ocean currents known as the *thermohaline circulation*, or the global oceanic conveyor belt. As cold, salty water sinks, more water moves north to replace it, thus speeding up the North Atlantic Drift and the Gulf Stream. They, in turn, are part of the global system of ocean currents both at the ocean surface and deep underwater; important parts of the earth's climate system.

Polar Talk

North Atlantic Deep Water is a mass of dense, cold, salty water that forms on the surface of the northern Atlantic, primarily in the Norwegian Sea, and descends to flow south deep under the ocean.

The vertical movement of ocean water driven by density differences is **thermohaline circulation.** The term is derived from the Greek words for "heat" and "salt."

Antarctic Water's Ups, Downs Are Important

If you voyaged south toward Antarctica on a ship, you could celebrate reaching the Antarctic when you crossed the *Antarctic Convergence,* long before reaching the Antarctic Circle. The Convergence is a water temperature, biological, and legal dividing line. When you cross it, the water and the air above it grows colder and you begin to find kinds of sea life not seen farther north. The Antarctic Treaty makes the Convergence the northern boundary of the zone where ocean life has special protections.

Polar Talk

The oceanic zone between about 50 and 60 degrees south where cold water from the Antarctic region meets and sinks beneath warm water from the middle latitudes is known as the **Antarctic Convergence,** sometimes known as the Antarctic polar front.

The exact location of the Convergence shifts with the winds and currents, but it's between latitudes 50 and 60 degrees south, or about 1,000 miles (1,500 kilometers) north of the Antarctic continent. It's where cold ocean water moving north converges with warm water flowing south, with denser cold water flowing under warm water.

As you continue across the ocean toward Antarctica, you encounter another zone where ocean water is moving vertically, but this time, it's coming up instead of going down. This happens in the zone called the *Antarctic Divergence.* The water that's coming up here is known as *circumpolar deep water.* This water began life at the other end of the earth, in the northern Atlantic, as North Atlantic deep water, but has been modified enough by other water to earn its new name. As we'll see in Chapter 7, this water is important to the rich sea life of the Southern Ocean.

Polar Talk

The oceanic zone just north of the Antarctic continent where west-flowing and east-flowing currents diverge is the **Antarctic Divergence,** which allows salty, nutrient-rich water to come to the surface.

Circumpolar deep water is North Atlantic Deep Water that has been modified by interactions with other water masses on its long journey south to Antarctica, where some of it upwells. (Upwelling is the process by which deep, cold, nutrient-laden water comes to the ocean's surface. This occurs in areas of divergence, such as around Antarctica or along the equator, or where winds blow water offshore, such as along the California coast.)

Wind and Earth's Rotation Create Divergence

As water is being pushed along as an ocean current, the earth is rotating under it. If you could see a particular "parcel" (imagine a box of water, now remove the box) of water, you would see that the wind and the water are traveling in a straight line in relation to space, that is, to the background of stars. But the wind and water are following a curved path over the rotating earth. In the 1830s, the French mathematician Gustave-Gaspard Coriolis worked out the mathematics of this; thus we talk about the Coriolis force causing winds and ocean currents to curve.

In the Southern Hemisphere, the Coriolis force causes ocean currents to turn toward the left. Winds generally blow from east to west close to the Antarctic continent, pushing the water to the west. It turns toward the south. Farther north, winds are blowing from west to east, pushing water to the east. It turns left, toward the north, creating the divergence. This turn toward the north also helps create the Antarctic convergence. Here, surface water meets water that was pushed toward the south when the east-to-west ocean currents in the tropics south of the equator turned toward to the left.

Crevasse Caution

The Coriolis force helps explain why winds blow in different directions around storms in the Northern and Southern hemispheres and why ocean currents bend different ways in the two hemispheres. It does not cause water to swirl in one direction or another while going down a drain. Coriolis affects only large wind and water movements, such as hurricanes or long-lasting ocean currents.

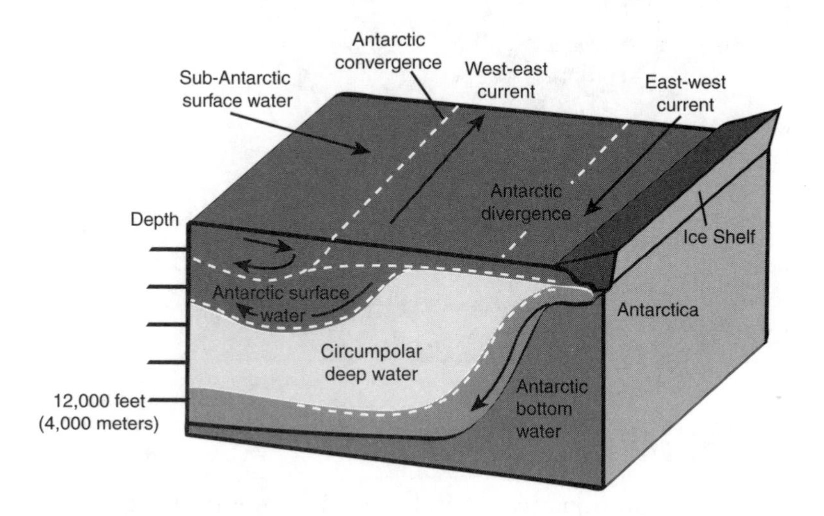

Movements of Southern Ocean water.

World's Largest Ocean Current Circles Antarctica

When the British explorer Captain James Cook sailed south toward Antarctica in 1773, he discovered the Southern Ocean has steady west-to-east winds, large swells, frequent storms and an ocean current that moves to the east. This *Antarctic Circumpolar Current* is the world's only ocean current that flows through all three major oceans: the Atlantic, Indian, and Pacific.

> **Polar Talk**
>
> The **Antarctic Circumpolar Current** is the world's largest ocean current, which moves all around Antarctica. It connects the waters of the Atlantic, Indian, and Pacific oceans.

For most of its trip around the world, the current runs free of land. The only restriction is the Drake Passage, the 600-mile (1,000-kilometer) gap between the northern tip of the Antarctic Peninsula and the islands at the southernmost part of South America. The current moves slowly, about 1.65 mph (2.75 kph), but it carries a lot of water; estimated at approximately 170 cubic yards (130 cubic meters) of water per second through the Drake Passage.

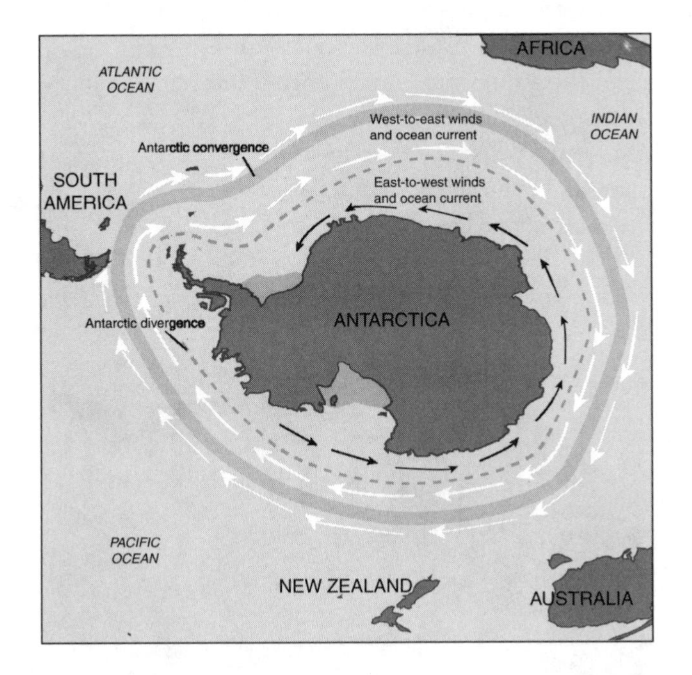

Antarctic ocean currents.

We Can't Escape Polar Ice

The next time you're enjoying a vacation at a tropical beach look out over the water and think about what's under the water besides the fish you expect to eat that night. If the water is deep, as it is a few miles off many tropical islands, you would find cold, salty, nutrient-rich *Antarctic bottom water*. Scientists estimate that water, which covers about 75 percent of the world's ocean bottoms, came from around Antarctica. It has a temperature of around 31°F (–0.5°C) and is about 3.47 percent salt.

Antarctic bottom water is created as ice forms, expelling salt, which sinks and makes the water under the new ice saltier. Strong winds blowing over the ocean from the Antarctic ice cap help form bottom water. Even during the coldest weather, these winds can blow ice floes apart, creating polynyas. As soon as a polynya opens, its water is exposed to bitter cold and quickly begins freezing, as more salt goes into the water.

Polar Talk

Antarctic bottom water is the cold, very salty water that forms around Antarctica and flows along the bottoms of all the world's oceans.

Like the cold, salty water that sinks in the North Atlantic, Antarctic bottom water is part of the system of ocean currents that move nutrients to parts of the ocean that would otherwise have little life. Maybe even more important, these currents help the earth balance its heat budget by bringing cool water to the tropical regions and moving warmth into the polar regions.

The Least You Need to Know

- Freezing and melting of sea ice around Antarctica each year is the world's biggest seasonal climate event.
- A layer of very salty water protects Arctic sea ice from being melted by warm water below it.
- Water flowing from the Arctic into the Atlantic ocean helps drive global ocean currents.
- The Antarctic Convergence is a water and air temperature, biological, and even legal boundary.
- The world's largest ocean current circles Antarctica, linking the Atlantic, Indian, and Pacific oceans.
- Cold, salty water sinking to the ocean bottom around Antarctica helps control global climate.

Part 2

Life in the Polar Regions

In our day-to-day lives, we use ice and cold air to keep food from spoiling; we don't think of them as providing environments for plants and animals to live and even to thrive. Yet as we'll see in this part, the frigid Arctic Ocean and the Southern Ocean around Antarctica are more than hospitable to life ranging from single-cell creatures you'd need a microscope to even see, up to the largest forms of animal life on Earth, whales. Even the ice that floats in these oceans is far from dead; it's a key to abundant life. In fact, parts of the polar oceans are some of the most "productive" on Earth at certain times of the year in terms of the total amount of plant and animal life inhabiting them.

Chapter 6

Arctic Ocean Life: Algae to Polar Bears

In This Chapter

- Ocean life begins with plankton
- Ice gives life a toehold
- The icy, dangerous life of seals
- Whales beat the cold with size
- Polar bears are the top predator

In this chapter, we will begin with the smallest plants and animals, and work our way up the ocean's food web, looking at a few of the many kinds of Arctic Ocean inhabitants, focusing on those you won't encounter at your local beach unless you live on the Arctic Coast. We'll meet tiny plants and animals, including some that find comfortable homes in the sea ice that floats on the Arctic Ocean. From here we'll move on to fish, seals, whales, and finally polar bears. Yes, polar bears. They—unlike any other kind of bear—are considered to be marine mammals because they obtain most of their food from the ocean.

Swimming Up the Arctic Ocean Food Web

The basic theme of life is that tiny animals eat tiny plants; bigger animals eat plants or smaller animals.

All life needs certain things. For plants, these are carbon dioxide, water, and small amounts of other minerals needed to produce the carbohydrates, sugars, and starches that plants are made of. Plants also need energy, which almost always comes from the sun, to turn the raw materials into the stuff of plants. Animals need energy and materials to build their tissues. Almost all Earth's animals obtain the energy and materials they need, either directly or indirectly, by eating plants that grow with the help of solar energy. Animals that eat other animals are tapping into a food web that begins with plants, because if you trace it far enough back, you'll find an animal that eats plants.

Polar Talk

Plankton is very small plants and animals that mostly drift with currents, although some are weak swimmers. Tiny plants that drift with the currents in water are **phytoplankton.** They are the source of most of the ocean's plant material. **Zooplankton** is animal plankton.

Polar Talk

Photosynthesis is a series of chemical reactions in plants that convert radiant energy from the sun into chemical energy that plants can use.

Most Oceanic Life Begins with Plankton

When we begin talking about an ocean's food web, we start with the two kinds of *plankton*, *phytoplankton* and *zooplankton*. Plankton include mostly microscopic algae, protozoa, and the larva that will grow into higher forms of animals, if nothing eats them.

Various kinds of algae directly or indirectly supply 99 percent of the food that marine animals eat. In other words, the oceans' phytoplankton is mostly algae. Algae use the energy of sunlight to produce their own food using *photosynthesis*. Zooplankton and many larger creatures, including humans, eat algae directly. The algae eaten by zooplankton is, in turn, eaten by tiny animals that are meals for slightly larger ones and on up. In other words, almost any food from the ocean, whether it's the wild salmon you eat or the seal that a polar bear grabs, can be traced back to algae.

Some Life Thrives in Cold Water

The cold water and even the ice of the polar oceans turn out to be friendly places for life at the bottom of the oceanic food web. As the temperature of water goes down, it can hold more dissolved gasses. The additional gasses in cold water helps ensure plenty of carbon dioxide for phytoplankton to use. When zooplankton eat phytoplankton, the extra oxygen dissolved in cold water helps ensure they won't be short of breath.

Tiny Life Finds a Home in Ice

As we saw in Chapter 5, when sea ice quickly freezes, it traps some salt, which creates pools of brine (salty water) inside the ice. As time goes on, the brine melts through the ice, forming a series of channels leading to the water the ice is floating in. As the ice is forming, it also traps bacteria, algae, and small creatures.

When the sun comes up in the spring and the ice begins to melt, sunlight reaches the algae, which begins growing, staining the ice with a greenish-brown color. The bottom of the ice becomes a banquet table for a host of tiny sea creatures, which not only find food, but also hiding places in the brine channels from larger creatures that want to eat them. All of them can't hide, which is why larger creatures come to the bottom of the ice to graze. Scientists estimate that sea ice in some areas could account for maybe half of the plant material formed as the basis of the Arctic Ocean food web.

Let's Go Up the Food Web

The ocean food web is complicated with scores of kinds of creatures finding their own niches. Some stick to the bottom of the ocean while others move around at different water depths. Most of us are familiar with fish, but not with the different creatures that fish eat. If we travel up the food web from phytoplankton to a fish we'd find microscopic creatures, worms, various kinds of crustaceans—such as shrimp, crabs, and very tiny *copepods*.

The Arctic cod is the Arctic Ocean's main fish, and they eat more Arctic plankton than any other creature. They are, in turn, eaten by sea birds, seals, and beluga whales. Arctic Cod live all around the ocean and have been seen as far north as 82 degrees, 42 minutes latitude, farther north than any other fish. They seem to be most comfortable in water from around 32°F to 39°F (0°C to 4°C). They grow to be about 12 inches (30 centimeters) long and apparently don't live longer than about 6 years.

Polar Talk

Copepods are microscopic to nearly microscopic crustaceans that are a common kind of zooplankton.

Crevasse Caution

Life is more like a food web than a chain. The California roll you enjoy at a sushi bar illustrates this. The fish was next to the top of a chain (you're at the top). It ate a little fish, which swallowed a shrimp, which gobbled down an almost microscopic crustacean, which nibbled algae. The roll's wrap is algae. If you had shrimp sushi, you fed on another part of a food web.

Keeping Warm While Living in Cold Water

Fish, such as the Arctic cod, which live in water where the temperature regularly drops to 29°F (–1.8°C) have a big problem: how to keep the water in their tissues from freezing. Fish are cold-blooded animals. This means they have no way of *thermoregulation*, that is, of keeping their bodies at an ideal temperature. The fish's body cools to the temperature of the water it's swimming in. Polar fish have evolved variations of one answer: They produce antifreeze that their blood carries around their bodies.

Polar Talk

The ability of an animal to maintain body temperature within a narrow range despite changes in the environment is **thermoregulation.** Animals that can do this are "warm blooded," animals that can't are "cold blooded."

Anyone who works out regularly knows that you don't just begin lifting weights, running, or seeing how high you can get on the StairMaster. You first warm up. You do this because muscles work best when they are a few degrees above their resting temperature. This is just one example of how cells and tissues are most efficient at an ideal temperature and, in fact, might not work at all if they grow too warm or cold.

Imagine what happens when a seal—a warm-blooded marine mammal—with muscles warmed up and brain sharp, encounters a sluggish, cold-blooded Arctic cod in near-freezing water. The seal will have lunch. The cod will be lunch.

Marine Mammals Adapt to a Cold Life

So far we've looked at cold-blooded creatures, such as fish. We'll spend the rest of this chapter on *marine mammals* that live in the Arctic in harmony with the coming and going of its sea ice. As warm-blooded mammals, seals (including their close relative the walrus), whales, and polar bears, have evolved ways of keeping warm in the Arctic. One of the most important of these is to grow a thick layer of fat under the skin. In marine mammals such as seals and whales, this is called blubber.

Polar Talk

A mammal is an animal that gives birth to live young, nurses its young, and maintains a constant body temperature. A **marine mammal** depends on the sea for all or almost all of its food.

Marine mammals breathe air, they do not have gills like fish that enable them to extract oxygen from water. Seals and walruses *haul out* onto the ice at various times for various reasons, including to give birth. Because they give birth to live young, marine mammals that come out of the water onto land or ice need to shelter their young from the cold and predators until they are mature enough to venture onto the ice and into the cold water.

As mammals, whales breathe air, but what would be a nose on other mammals is a blowhole on the tops of their heads. To breathe, a whale can keep most of his body in the water with only the blowhole exposed. Whales, of course, give birth to live young, but because whales never leave the water, their young are born with the ability to stand the water's chill.

Polar Talk

In reference to a marine mammal such as a seal, **haul out** or **hauled out** is to come out of the water, to be out of the water on ice or land.

A Seal's World: Ice, Cold Water, Danger

Ten species of seals live around the Arctic, with the ringed seal being the most common. Our brief look at a year in the life of a ringed seal will help you understand a little about the lives of all seals around the Arctic.

From mid-March through early April each year, seal pups are born into a dangerous world. Ice is the best protection for a seal; when a seal is under the ice she's safe from polar bears. But seals have to come up to breathe, and to give birth. For this, the mother-to-be hollows out a den in a snowdrift on the ice, with a hole inside the den that opens to the water below the ice. Pups are born with a white, wool-like coat, which helps keep them warm until they grow enough blubber after five to eight weeks of nursing.

The den shields the pup from the wind and—most of the time—hides pup and nursing mother from predators such as polar bears. Pups learn to dive while very young. If the mother senses danger, both can dive through the hole into the water. Still, at any moment the lives of pup and mother could end as 1,200 pounds (545 kilograms) of polar bear, which had caught a whiff of seal, comes crashing through the roof. When the ice begins breaking up, usually by July, the pup is on his own as mother swims off to live her own life.

In May and June, you're likely to see ringed seals on the ice, basking in the sun as they *molt*. This probably helps the new hair grow. While they might be basking in the sun, the seals are alert, raising their heads every 20 seconds or so and quickly slipping into the water if they see any sign of a polar bear or human.

Polar Talk

To shed feathers, hair, or skin periodically, often with the season, is to **molt**, which allows new feathers, hair, or skin to grow.

As the ice begins to grow in the fall, ringed seals begin cutting breathing holes in the ice, using the sharp claws on their forelegs. Because they keep these holes open all winter, even if the ice grows to around 7 feet (2 meters) thick, ringed seals can live in areas the ice completely covers.

Seals dive for food, but because they are mammals, they have to come up for air every 5 to 15 minutes. If a seal had only one or two breathing holes in the ice, it would be easy prey for a patient polar bear. Instead, a seal might have 15 or so breathing holes, which greatly improves a seal's odds of not meeting a hungry polar bear when it comes up for air.

All Whales Are Big, but Not Alike

If you saw a narwhal for the first time without knowing what to expect, you might wonder why anyone would go to so much trouble to play a practical joke, especially in the Arctic Ocean. You'd see a light gray, almost white, whale maybe 15 feet (5 meters) long, but it has a 6-foot-long (2 meters) spiral tusk coming from its head. You could wonder: *Is this real?* It is.

A narwhal's tusk is really a tooth, but that's about all that scientists can say about it with certainty. No one really knows what narwhals do with the tusk. People have speculated that it's used to spear fish, but if this is the case, how would they get a fish off the tusk? Scars on male narwhals raise the possibility that they use the tusks to fight when they're seeking mates.

As whales go, narwhals are on the small side. But they are still large animals and this illustrates one evolutionary strategy for keeping warm: Decrease the surface to mass ratio. Heat is lost through an animal's skin that's exposed to the air. The more skin there is in relation to the amount of stuff inside the skin—bones, muscles, fat, organs—the faster heat will be lost.

For land animals, there's one huge problem with becoming bigger: The legs have to be stout enough to support the weight. But any creature that floats in the ocean obviously doesn't need legs to support it. In effect, if something is floating, it's weightless. This is why the biggest dinosaurs were never as large as blue whales—which can be 100 feet (30 meters) long. Being weightless in the water gives whales another advantage, they can carry around a load of fat—blubber—that a land animal couldn't afford.

Big Whales Feed on Tiny Creatures

Narwhals are one of only three species of whales that spend their entire lives in the Arctic. The other two are the beluga and the bowhead. Both the narwhal and beluga are "toothed whales" that eat things such as fish, squid, and other creatures fairly far up the food web.

The third whale species that lives in the Arctic, the bowhead, is the other group, the "toothless" or baleen whales. The world's biggest animals, such as the blue whale, are baleen whales, which feed by sucking in plankton through bony plates—the baleen—in their mouths. Even though baleen has traditionally been called "whalebone," it's not bone, but is made mostly of keratin, which is also found in human hair and fingernails.

To eat, bowheads swim with their mouths open to strain zooplankton from the water with fine fringe along the edges of their baleen, which hangs down somewhat like the teeth of a comb from both sides of their upper jaws. The more than 600 baleen plates are as long as 13 feet (4 meters). They feed near the surface, at the bottom of the ocean, and under ice.

You Don't Want to Meet a Polar Bear

Arctic Ocean ice researchers are among the few, if not the only, scientists who are routinely accompanied by someone with a rifle when they go to work. The person with the rifle isn't there to shoot a seal for dinner, but in case a polar bear decides to see if a scientist tastes as good as a seal. Polar bears are among the very few North American animals that will attack humans without being backed into a corner. Fortunately, relatively few humans venture into polar bear territory and almost all those who do know how to avoid them.

Although polar bears are considered to be marine mammals, they spend most of their time out of the water, but hunt mostly other marine mammals, especially seals. They are strong swimmers, however, and easily cross open water to find better hunting. Their white fur, well-developed sense of smell, extremely sharp claws, strength, speed, and skill at patient stalking make them dangerous to anything else that ventures onto the ice or into leads through the ice. They even kill and eat young whales that come too close to the ice edge.

Except for mothers with cubs, polar bears generally roam the ice alone for most of the year. In the spring, males begin looking for females by following their tracks across the snow atop the ice. Males and females also tend to congregate in good seal hunting areas. In the fall, a pregnant female finds a good area to excavate a den under snow or ice either on land or on the sea ice. Two, or sometimes three, very small, almost helpless cubs are born in December and spend the first three or four months of their lives in the den, nursing from their mother, who is living off her fat reserves. After emerging from the den in the spring, the cubs will spend a couple years with their mother before going off on their own.

U.S. law no longer allows polar bears to be hunted for trophies like this one in the lobby of the Hilton Hotel in downtown Anchorage, Alaska.

(Darlene Shields)

Unlike other bears that slow their metabolism to hibernate in the winter, polar bears can slow down their metabolism any time food becomes scarce. This means that in areas such as Hudson Bay, where most of the pack ice melts in the summer, polar bears move onto land, where they are mostly inactive, living on stored fat, and grabbing whatever food is handy, including even grasses and berries.

Obviously, all the animals that live in the Arctic Ocean or on the land surrounding it have adapted well to a harsh climate. In a sense, polar bears have adapted the best. Until humans began moving into the Arctic, polar bears had nothing to fear.

The Least You Need to Know

- Some kinds of life thrive in cold water and even ice.
- Fish that live in cold water have evolved antifreeze to keep water in their tissues and blood from freezing.
- Marine mammals keep their body temperatures in a narrow range with layers of fat and other adaptations.
- Arctic seals live in a harsh world of ice and stalking polar bears.
- Whales' large size is an adaptation to life in cold water.
- Large baleen whales feed on some of the ocean's smallest creatures.

Chapter 7

Life Adapts to the Frigid Southern Ocean

In This Chapter

- Underwater currents and ice create a unique environment
- Krill are key to ocean life
- Fish are unlike most others in the oceans
- The environment agrees with seals
- Whales come south to eat

The Southern Ocean is much more hospitable to life than the mostly ice-covered Antarctic continent that it surrounds. In this chapter, we'll first look at how global ocean patterns and the unique nature of the icy Southern Ocean combine to encourage life. Then we'll turn to some Southern Ocean creatures, including krill, a tiny shrimplike crustacean that many larger animals eat. Our journey around the ocean's food web will end with whales.

Regular Delivery of Nutrients Enhances Life

In Chapter 5, we saw how water sinks in the North Atlantic and flows deep under the ocean toward the south, mixing with water from other sources. Eventually, some of it comes to the surface around Antarctica, where oceanographers generally call it circumpolar deep water.

During the water's slow journey to the south, organic matter, including waste from animals and little bits of stuff that was once alive, slowly rain down from the water above. At the top of the ocean, where the sun shines, algae recycles such organic matter into new plant material. Large amounts of organic matter, however, fall into the ocean's dark depths where photosynthesis can't recycle it back into life. This material becomes increasingly concentrated in deep-ocean water.

Polar Talk

Nutrients are organic compounds from living things such as proteins and carbohydrates, and inorganic compounds, which are not from living things, such as nitrogen and phosphorous, that are necessary for plants and algae to grow.

When *nutrient*-rich water comes to the surface around Antarctica, it can play a role like that of rich fertilizer spread on a field. Life could bloom except that, depending on how far south of the Antarctic Circle you are, the sun doesn't rise over the ocean for weeks, maybe two months in winter. Summer's endless days help make up for winter darkness, but all life has to survive the long polar night.

Ice Helps Create a Buffet of Krill

In Chapter 6, we saw how Arctic Ocean sea ice helps algae, the base of the food web, hang on during the winter. A similar thing happens in the Southern Ocean. As in the Arctic, when the sun comes up, algae, which has spent the winter in the ice, explodes into growth to support complex communities of life.

In the Southern Ocean, the most common creature you'll find at ice's banquet table or floating in the water when the ice melts is a small crustacean known as *krill.*

Polar Talk

Krill is the common name used for members of the crustacean order *Euphausiacea.* Five of the world's 80 plus species are found in the Southern Ocean.

Female krill are believed to lay 8,000 to 10,000 eggs at a time, more than once a year. Many of these are eaten before they grow into adults, but huge numbers manage to grow up to roam the Southern Ocean in schools that can stretch for miles with 30,000 in a cubic yard.

All crustaceans molt as they grow. That is, they shed their old shell when it begins growing too tight, and grow a new one. Krill, however, have a strange twist on this. At the end of the summer, adult krill lose their sexual characteristics, go through a series of molts, and end up looking like two-year-old juveniles. In the spring, they begin to grow into adults again with a series of molts before spawning.

Turning back into juveniles is almost surely part of their adaptation to surviving during the long, dark winter when phytoplankton isn't growing. Unlike many other animals, krill don't build up fat reserves to carry them through a winter. Laboratory tests have shown that krill can go as long as 200 days without eating. Using their own body protein instead of fat to supply the energy they need during their hungry time is part of their survival story. There's also evidence that krill might find food in the winter. Why would anyone care about krill? Krill are important because most of the larger Antarctic animals, the seals, whales, and seabirds as well as the less glamorous fish and squid, either eat krill or eat other creatures that dine on krill.

Antifreeze Is Only Part of Survival

Like some Arctic Ocean fish, especially the family of fish known as *notothenioids*, many in the Southern Ocean have antifreeze to keep their blood from freezing. This isn't enough, however, for fish to thrive in cold water. As the temperature drops, blood's *viscosity* increases. It resists flowing through arteries and veins. Most Antarctic fish have fewer red blood cells than fish elsewhere. This thins the blood, which allows freer flow, but it also reduces the amount of oxygen the blood carries since the hemoglobin that makes red cells red carries oxygen.

Even with this disadvantage, *icefish* have taken the idea of fewer red blood cells to what, at first, might seem a ridiculous extreme: They have no red blood cells. Obviously, it works, but it is unique. Antarctic icefish are the only known *vertebrates* that do not have red blood cells.

Needed oxygen reaches icefish cells by being dissolved in the blood, not attached to red blood cells. Here, the cold helps. As a fluid cools the amount of a dissolved gas it can carry increases. In addition, icefish have more blood than other animals of their size, and their arteries, veins, and hearts are larger than you would expect given their size. The name "icefish" comes from their pale, bloodless color.

Polar Talk

The property of a fluid that resists flowing is **viscosity.** In most fluids, it increases as temperature falls. **Icefish** are Antarctic fish that are members of the family *Channichthyidae* with no hemoglobin. An animal with a segmented spinal column and a well-developed brain is a **vertebrate.** This includes mammals, birds, reptiles, amphibians, and fish.

Marine Mammals Thrive in the Southern Ocean

While the Southern Ocean has some unusual fish, such as the icefish, not found in the Arctic Ocean, the same kinds of marine mammals live around Antarctica, with one, big, important exception. No polar bears live on or near Antarctica. Polar bears evolved from bears in warmer climates. To reach the Arctic, bears had to walk from the forests of middle latitude parts of Asia, Europe, and North America. No creature can walk to Antarctica.

Crevasse Caution

"Summer" and "winter" in Antarctica aren't like the seasons elsewhere, there's no real spring and fall because the seasonal shifts are so swift. Around the coast, in places such as McMurdo, the average low temperature is above 0°F (–18°C) only in November, December, January, and February. In this chapter and elsewhere in this book, "summer" in Antarctica refers to these months.

Seals Are at Home in Antarctica

The combination of an ocean with plenty to eat, plus ice and land for *pinnipeds* to haul out on, makes Antarctica and the Southern Ocean's islands the world's most hospitable place for seals. Six of the world's thirty-five species of seals live in the Southern Ocean, and scientists estimate that they make up more than half of the world's *biomass* of seals. Elephant seals and fur seals live mainly around the islands away from the Antarctic continent. The other four species—crabeater seals, leopard seals, Ross seals, and Weddell seals—are found on the ice around the continent.

Polar Talk

Pinnipeds is the suborder of marine mammals that includes seals, sea lions, and the walrus. **Biomass** is the total mass of a particular organism or group of organisms living in a location or the entire world.

Crabeater Seals Really Chow Down on Krill

The early explorers who gave crabeater seals their name got it wrong. (No one seems to know how this happened.) Crabeater seals don't eat crabs because few, if any, crabs live around Antarctica. In a way this is a shame, an inaccurate name is attached to what is probably the most numerous species of pinnipeds on the globe. Just how numerous is an open question. Estimates of the total number of crabeater seals range from around 12 million to more than 50 million.

A crabeater seal eats by swimming through large numbers of krill and sucking them in. It then squeezes the water out through its teeth, which have serrations that act like a sieve. This allows the water to escape while the krill stay in the seal's month. Research with captive crabeater seals has shown they can suck small fish into their mouths from 20 inches (50 centimeters) away. This ability probably enables crabeater seals to suck krill out of underwater crevices during winter when they are hard to find.

In September, a pregnant crabeater seal takes over a spot on the ice where she gives birth to a single pup. A male—which might or might not be the pup's father—joins the female and defends her and the pup. After the pup is weaned in about three weeks, the male and female mate, and the new pup will be born the next September. As the end of summer approaches and the water begins to freeze, most crabeaters begin heading north. Most seem to travel to the islands around the Southern Ocean, Australia, South America, and even South Africa.

Leopard Seals Spread Fear Across Ice

If you were an image consultant hired by the pinnipeds, your first advice would be to "get rid of the leopard seals. These guys trash the image of seals as cute creatures with big eyes." As an example, a World Wide Web advertisement for a documentary on Antarctica says: "Welcome to the icy realm of the leopard seal, a vicious predator with a huge, reptile-like head, gaping jaw and bladed teeth. They feed on seals and penguins and are known to attack humans …" Divers who conduct research in the ocean around Antarctica watch out for leopard seals instead of sharks, since sharks have never been seen in the ocean close to the continent. No record of a leopard seal killing a human can be found, but tales of their viciousness are among the first told to newcomers to Antarctica.

What a leopard seals does with a penguin it catches makes you want to avoid the same fate. The seal holds the penguin in its mouth, slapping it against the water until the penguin's skin peels off, allowing the seal to eat the featherless carcass.

Biologists don't know a lot about leopard seals. Females apparently give birth to single pups on the ice in November. The animals don't live in large groups and seem to wander around the Southern Ocean, living not only around Antarctica, but also visiting or living around other Southern Ocean Islands, on the southern Australian Coast, New Zealand, and the Atlantic Coast of South America. Leopard seals probably live about a quarter of a century, and their only natural predators are killer whales.

Whales Are Mostly Summer Visitors to Antarctica

Eight species of whales are regular visitors to the Southern Ocean close to Antarctica, but only the orca, or killer whale, seems to live there all year. Orcas, along with sperm whales, are the only toothed whales seen in the Southern Ocean near Antarctica. The other six species—the blue, fin, humpback, minke, southern right, and sei—are all baleen whales, like the bowhead whales we looked at in Chapter 6. All except the orca migrate to the Southern Ocean to spend the summer eating and then migrate to warmer water to give birth and nurse their young. During their time away from Antarctica, they eat little, living off the fat they put on in the Southern Ocean.

Huge Blue Whales Gobble Krill by the Ton

Blue whales are the largest animals on Earth today and are probably the largest mammal ever seen. Such a large animal obviously eats a lot. Blue whales are believed to feed almost exclusively on krill. The best estimate is that during its summer stay near Antarctica, a blue whale eats around 4 tons (3.6 tonnes) of krill a day. This works out to something like 40 million individual krill a day.

A blue whale female makes a huge energy investment in nursing a calf, which is probably why they breed only every two or three years. A newborn calf will be 23 to 27 feet (7 to 8.2 meters) long and weigh 3 tons (2,722 kilograms). It nurses for 7 to 8 months, consuming around 100 gallons (379 liters) of milk that's very rich in fat each day. A nursing blue whale gains about 8 pounds (3.6 kilograms) an hour. Blue whales seem to travel in pairs most often, but are sometimes seen in groups of 50 or more.

Orcas Are Southern Ocean's Top Predator

If you ever go on an Antarctic cruise, or visit McMurdo Sound, one of the highlights of the trip could be seeing the front several feet of a killer whale pop out of the water, almost straight up and down. Watching such *spy hopping* is part of the fun at a place such as Sea World, where orcas are a big part of the show. Seeing an orca pop out of the water of McMurdo Sound, with the Transantarctic Mountains across the sound in the background, is an experience. It connects you with the natural world in a way that a show in California or Texas or Florida never could.

> **Polar Talk**
>
> Whale behavior that refers to the animal thrusting its head straight up out of the water is known as **spy hopping**.

At Sea World, the killer whale is spy hopping because his trainer will reward him with a fish. In McMurdo Sound as elsewhere in the ocean, the orca is spy hopping to take a look around, to scratch itchy skin, or maybe to perform what scientists call a "display of dominance." That is, the killer whale is telling his fellow whales, "I'm the boss around here." Different killer whales have different places in the hierarchy of their own pods, or groups. In the larger world of Southern Ocean predators, killer whales are the bosses. They are the only animals that leopard seals have to worry about, since humans don't hunt leopard seals.

The Family That Hunts Together Eats More

Extensive studies of killer whales around the world have shown that they travel in groups of various sizes, beginning with a mother and calf, ranging through "clans" of a few whales up to "pods" of maybe 50 or more. Orcas make a variety of sounds including clicks, grunts, whistles, and squeaks. Individual whales have their own calls and whale researchers recognize "dialects" used by different groups of killer whales. The sounds apparently help pods coordinate hunting, but killer whales sometimes also hunt alone.

Although orcas live in oceans around the world, they seem to prefer the Arctic and Antarctic oceans. In the Southern Ocean, they seem to use hunting techniques that work elsewhere, but also some methods that wouldn't work in places such as off the California Coast. Around Antarctica, killer whales have been seen to throw their weight onto an ice floe, upsetting the floe and sending a penguin or seal sliding into the water for the orca to eat.

A few orcas have been seen in the sea ice around Antarctica in the winter, and these sightings have included small calves. This shows that some orcas breed near the Antarctic Coast. In fact, some scientists think that a separate species or maybe two separate species, as well as the one known elsewhere in the world, might live around Antarctica.

The Least You Need to Know

- Water from deep under the ocean and ice help algae blossom around Antarctica.
- All animal life in the ocean around Antarctica feeds on krill, either directly or indirectly.
- Antarctic fish have evolved antifreeze to keep their blood from freezing.
- More seals live around Antarctica than anywhere else in the world.
- Seven species of whales migrate to the Antarctic Coast to feed, but have their young elsewhere.
- Orcas, or killer whales, seem to be the only species of whale that breeds around Antarctica.

Chapter 8

Life Is Hard for Arctic Plants and Animals

In This Chapter

- Arctic plants are low hardy midgets
- Lemming population swings set the tempo for Arctic life
- A 10,000-mile trip with a tiny bird
- The ups and downs of musk oxen
- The secret difference between reindeer and caribou

In summer, the Arctic's land is cold and windy. In winter, when night falls, the land freezes solid. Yet plants and animals abound: those few species that have adapted to Arctic extremes. The permafrost, just a few feet down, nourishes no life. The tundra above supports plants that live close to the ground.

Land animals, like plants, of course, have adapted to the cold and wind, but in a wider variety of ways than plants, by evolving a variety of strategies for staying warm and obtaining and conserving food energy.

Plants Hug the Ground to Survive

No trees grow here, except along great rivers. Plants can't push roots into frozen ground and frozen cells die. During winter, plants snooze under snow blankets, awaiting the few weeks in the Arctic summer when the tundra thaws. As we saw in Chapter 3, short plants can grow in the active layer of the terrain. Other plants—lichens—survive on bare rock.

The harsh climate cripples or kills weak plants: low temperatures, continuous daylight in summer, continuous night in the winter, infertile soil that heaves and buckles with freezing and thawing, permafrost, strong dry wind, and blowing snow. Few species survive these conditions. Those that do are rugged dwarfs whose shallow roots skim the top of the permafrost.

The Arctic has more than 15,000 different kinds of lichen and 400 species of flowering plants: blue-spiked lupine, wild crocus, mountain avens, Arctic poppy, and saxifrage—tough midgets all. Arctic plants grow fast, maturing in record time. They jump-start in the spring when the snow still covers the land. In six weeks, they flower and seed: using 24-hours-a-day energy pouring from the sun.

Ice Chips

Lichens are strange brown, black, or gray, crusty, nonflowering plants that grow even on bare rock. They are composite organisms made of members of as many as three plant kingdoms: a fungus (the dominant partner), algae, and cyanobacteria (formerly called blue-green algae). The fungus is a "farmer" that can't make food. It cultivates the other partners that can make food and then the farmer eats what they produce. The fungus can, however, store water, which the others need. Hence the fungus farmer waters his crops. Lichens can shut down their *metabolism* and survive long periods, barely alive.

In July, the tundra flowers with brilliant colors: yellows, pinks, blues, purples, and reds. By late August, it's over, growing ends, and plants await their blanket of snow. Life sleeps in compact mounds, tough stems, roots, seeds, and spores.

Polar Talk

Metabolism is the complex processes that enable an organism to maintain life.

Tundra plants can grow at cooler temperatures (27°F to 36°F (–3°C to 2 °C) than any other plants on Earth. The low plants bask in the heat of the dark, warmed soil. Most plants are dark to absorb heat and hairy to keep it. Some plants grow in clumps to break Arctic winds and protect each other. Other plants sprout dishlike flowers that track the sun.

Animal Survival Means Not Being Eaten

Survival is a simple matter of getting enough energy and water and not being eaten. Not easy, though, where food is scarce, water is frozen, and predators lurk. The winners (among mammals) include caribou, musk oxen, Arctic wolf, Arctic fox, Arctic weasel, Arctic hare, and *lemmings*.

Animals Find Ways to Hoard Heat

To survive, animals must hoard heat. Mammals grow thick winter fur in two layers: a soft, dense underfur like a wooly blanket, covered by an outer layer of long, slick hairs that shed water, snow, and wind. Caribou and polar bears develop hollow hair for extra insulation and buoyancy. Birds grow lush, downy feathers. As we saw in Chapter 6, growing big is a defense against cold. For the same reason, animals evolved round shapes. Anything that sticks out—ears, snout, or legs—can freeze. So Arctic hares, for example, have short ears close to their bodies.

Polar Talk

A **lemming** is a mouse-size animal that looks like a miniature guinea pig.

Specialized Tools and Coverings Aid Survival

Over the eons, species developed special tools and coverings to survive. Floundering in snow consumes energy so hares have large feet like snowshoes. Caribou hooves flex to grip uneven tundra and spread for snow travel or river swims. In autumn, the front feet of collared lemmings develop two huge claws to shovel into tundra snow. Arctic foxes grow long hairs on the soles of their feet to give them traction on ice and extra insulation.

Some animals camouflage, becoming white in winter and brown in summer. Arctic hares, *ptarmigans*, Arctic foxes, collared lemmings, and Arctic weasels all change color and are called "varying" hares, etc., for this reason.

Polar Talk

A **ptarmigan** is a grouse that has feathered legs and feet for Arctic life.

Some Spend Winter in Deep Freeze

A few animals survive by freezing during winter, then thawing in the spring. It is a dangerous expedient. Freezing usually ruptures cells, killing the animal. However, wood frogs, wooly bear caterpillars, and a few insects in the Arctic can freeze successfully. More than half their total body water freezes between the cells. The cells themselves contain natural antifreeze so they don't freeze and burst.

Some Head South for the Winter

Ice Chips

The rust-colored wooly bear caterpillar of Ellesmere Island actively feeds for only about one month in the summer and freezes during winter. The thumb-size insect takes 14 years to reach adulthood and can withstand cold as low as –95°F (–70°C).

Some birds and animals migrate to find food and escape cold. The Arctic tern follows summer by flying 11,000 miles between the Arctic and Antarctic each year. Caribou herds shuttle hundreds of miles between the tundra and tree lands.

Others live under the insulating snow. Lemmings almost never appear on the surface during winter. Hares shelter under shallow pits in the snow and also sit atop snowdrifts on the lee side of boulders. Snow is good insulation; it keeps animals such as lemmings warm enough to survive the winter.

Moving Up the Food Web

Lemmings control the tempo of life in the Arctic by swinging their population from high to low numbers about every four years. Other species dance to their rhythm. Nearly every *carnivore* in the Arctic dines on these small rodents, about the size of a mouse. Consequently, when lemmings crash, predators die, especially the young.

Few if any snowy owlets and Arctic fox pups live during times of lemming dearth. Only pups born during a year of lemming abundance survive in sufficient numbers to sustain the fox population. Foxes dwindle until the next good lemming year.

Nobody knows why lemming numbers fluctuate so. What's more, it isn't a local phenomenon. The numbers are high or low frequently over a large area at the same time. Perhaps lemmings weed out their weak. When the population nosedives, the numbers hover scarily close to zero. It approaches species extinction. Yet the strongest lemmings survive and the species continues.

Polar Talk

A **carnivore** is a flesh-eating animal.

Lemmings are aggressive little creatures that fight as their numbers increase. The stress of overcrowding may change hormone levels, which can decrease birth rates. Also, with enough population stress, lemmings kill each other.

CAUTION

Crevasse Caution

Various theories about lemmings have been proven false or, at least, remain unproven:

- They commit mass suicide. There are no authentic accounts of suicides.
- Predators overeat lemmings. No, instead, lemmings control predator population.
- Epidemic diseases sweep through lemmings. Nope, virtually no disease occurs during some declines.
- Lemming food runs short: Food does vary but nobody has been able to show a cause and effect.

Arctic foxes prowl the tundra for food, including even the smallest and most remote islands north of Canada and Greenland (they sail on ice floes to reach these places). Arctic foxes roam within 300 miles (485 kilometers) of the North Pole.

They will eat almost anything and follow polar bears to scavenge the way jackals follow lions. The bears eat only blubber from seal kills; the foxes gobble the rest. In the summer, when food is abundant, foxes kill more than they can eat to cache the excess in their dens under stones and in crannies for lean winter months. One such store held 50 lemmings and 30 to 40 auks, each lined up side by side, with each head neatly bitten off and tails pointed in the same direction—a tidy mind at work.

Birds Flock to the Arctic in Summer

In late May and early June, birds flock to the Arctic eager to gobble its insect swarms and fat cotton grasses. More than 180 bird species breed in the Arctic every year and then head south. Bird families include ducks, geese, shore birds, jaegers, gulls, terns, warblers, loons, and swans. Only a few hardy ones reside the year round: gyrfalcon, ptarmigan, raven, the little auk, and the snowy owl.

Snowy Owls Are Polar Wanderers

His huge wings—spread as curved fans, each distinct feather backlit by the sun—break his descent in absolute silence. The snowy owl alights on a cliff's edge with an Arctic hare to feed his hungry brood. The chicks look like scattered, small, dark rocks on the cliff top, some larger than others due to staggered hatching. Their nest is a small depression in the ground with a panoramic view. Would-be predators stand little chance of sneaking up undetected. Normally still as stones, the chicks break pose to beg food from their father, nibbling at his bill and feathers, as the mother takes off to hunt.

A snowy owl chick caught on the tundra near Barrow.

(Jack Williams, USATODAY.com)

Among the strongest of owls, although not the largest, snowy owls wander all over the Arctic. They range from Ellesmere Island to the southern shores of Hudson Bay and from northern Siberia to the Shetland Islands off the coast of Scotland. They follow food and leave the Arctic only when lemmings disappear. Then they've been sighted as far south as the northern 48 contiguous U.S. states.

Birds Stay Busy in the Short Summer

In late May to early June, birds swoop into the Arctic, stake out territory, and display and sing advertisements for mates and warnings to trespassing males. Late June finds most females sitting on nests, incubating eggs, and trying to look inconspicuous.

Males take off for lagoons remote from prowling Arctic foxes to molt and grow new flight feathers. Flightless and practically helpless while molting, they feed and rest in safe harbors.

Meanwhile, chicks peck egg shells and a thousand pecks later breathe outside air. By mid-July most have hatched, and by mid-August they're ready to fly. In late August and early September, they take off south for winter grounds. Arctic migrants are a worldly bunch: Dunlins take off for China, northern wheatears for Africa, Arctic terns for Antarctica, and sandpipers for Argentina.

Don't Mess with Musk Oxen

What could be a better defense? Snow and wind buffet their broad shoulders. They face outward—the bulls—with weapons ready, guarding the cows and calves huddled in the center. Gray shapes slip in and out of the blowing snow. One attacks. A bull hooks a curved white horn into the flank, drives broad hoofs down, and tramples the wolf beneath 700 pounds (320 kilograms) of body mass.

More than 1,000,000 musk oxen roamed the Arctic tundra during the ice age of the *Pleistocene Epoch*. Then men with guns arrived. The musk oxen stood defenseless against such an enemy. Bulls, cows, and calves were easy shots: just standing there, a motionless closely grouped target. In 1894, an *Inuit* may have shot the last Alaskan musk ox and thought it was "a bear with horns." By 1900, the musk ox was rare in Canada and nonexistent in Alaska.

Polar Talk

The **Pleistocene epoch** is a geological period 1.6 million to 10,000 years ago. During that time, ice covered northern areas and then retreated.

The **Inuit** are native people that live in the Arctic regions of North America, especially Canada and Greenland. The native peoples of Canada and Greenland generally prefer to be called Inuit and those of Alaska, Eskimo.

In May 1930, the U.S. Congress appropriated funds to bring musk oxen back to Alaska and told the old U.S. Biological Survey "to acquire a herd of musk oxen from Greenland for introduction into Alaska with a view to their domestication and utilization in the Territory." Later that year, 19 female and 15 male musk oxen were transported to Alaska. Now 3,500 of the ice-age creatures wander Alaska from the Yukon River delta in the southwest to the Arctic National Wildlife Refuge in the northeast.

Caribou Are Born Free

About 2 million caribou roam the tundra and forests of North America from Alaska to Labrador. Some herds migrate 3,000 miles (5,000 kilometers) a year—farther than any other land animal. Other herds travel shorter distances and some (the woodland caribou) don't migrate at all or no more than about 40 miles (65 kilometers).

In general, caribou make two major migrations a year: one for food, the other for shelter. In the spring, they seek the lush tundra, greening to the north. Cows head there to drop their calves and bulls follow soon. In the fall, they retreat south to the forests for winter shelter.

During winter, the Porcupine Caribou Herd browses in snowy forests south of the Brooks Range. In April, a cow decides to go north to calve. Other cows follow, moving at a brisk walk, in single file. The bulls trail in a few weeks.

Ice Chips

Caribou cows give birth the first week of June, almost simultaneously—the whole herd. About half drop their calves within days and 90 percent within a fortnight of each other.

Nobody knows which route they will travel from year to year. They'll probably use the well-beaten route—swimming rivers and contending with wolves—to calving grounds on the Arctic coast, about 400 miles (645 kilometers) away. But maybe not. In addition to the Porcupine Herd, the Central Arctic Herd also migrates to the Arctic Ocean into the Kuparuk and Prudhoe Bay oil fields, where the females calve.

In late June and early July, the whine of freshly hatched mosquitoes and the buzz of flies fill caribou hearts with dread. They move together in the tens of thousands to the coast, to ice fields, to uplands in the Brooks Range—anywhere to escape the torment. By mid-July, the herd heads out, east and south, back to the forests to spend the winter.

Caribou and reindeer are the same species but they sure act differently. The Sami (Lapps of Scandinavia and eastern Siberia) tamed the reindeer centuries ago. Consequently, reindeer don't migrate, but they do pull laden sleds.

In the late 1800s, the Reverend Sheldon Jackson brought reindeer from Siberia to the Seward Peninsula of Alaska. In 1937, the U.S. government transferred ownership of the reindeer (now numbering in the hundreds of thousands) to Alaskan Natives.

Ice Chips

In the 1990s, times were good for caribou herders, who were getting $50 a pound for reindeer velvet to cure sexual impotence. Then the price dropped to $10 when Viagra hit the market.

In November 1996, Tom Gray, a reindeer herder, lost 700 reindeer in a single day. Thousands of migrating caribou wandered through the snowy tundra hills where his peaceful herd of 1,100 dwelled about 60 miles (100 kilometers) east of Nome. The old wild urge to migrate overcame more than half of Gray's herd. The reindeer moved out with the caribou, turning wild.

Caribou lured such numbers from the Steward Peninsula that 8 out of 15 Inuit reindeer herders went out of business. Gray keeps his remaining few isolated from the wild ones on a strip of land that juts into Golovin Bay. It's not much of a buffer, the caribou are only six miles away.

The Least You Need to Know

- The Arctic has only a few species but many animals.
- Tough midget plants survive, rooting in soil that thaws a few inches deep in the summer.
- Arctic foxes and snowy owls eat anything and roam far to find it.
- Almost 200 bird species flock to the Arctic each year, some traveling thousands of miles.
- Hunters wiped out the musk oxen in Alaska, but the U.S. government restored the herd.
- Caribou and reindeer are the same species, but caribou migrate freely, some going farther than any other mammal.

Chapter 9

Living on the Antarctic Continent Is Tough

In This Chapter

- Dryness, not just cold, makes Antarctic life difficult
- Land animals are almost too small to see
- Microbes live in cold, deep ice
- The world's largest bird breeds in the Antarctic
- Those adorable penguins live a hard life

Because all but about 2 percent of the Antarctic continent is ice, plant life, the basis of the food web on land, has hardly any hope of gaining a toe-hold. To make it even tougher, large parts of the ice-free areas are deserts, such as the Dry Valleys. Despite these challenges, a few plants manage to live in Antarctica and some tiny animals are around to eat them. In this chapter, we'll see how they do it. We'll also look at Antarctica's birds, especially the penguins, which are sea creatures in the sense that they find their food in the ocean. On the other hand, they nest on Antarctica's land. We will learn how they manage to survive on the world's harshest continent.

Cold and Dryness Are the Main Challenges

As we saw in Chapter 8, a wide variety of plants and animals live in the Arctic, but few plants or animals live in Antarctica. G. E. Fogg, in his book *The Biology of Polar Habitats*, sums up the differences: About 900 species of *vascular plants* live in the Arctic, but only 2 in Antarctica. The Arctic has 48 species of mammals. Antarctica has none.

Cold, obviously, is a barrier to life. But plants and animals have evolved various strategies for coping with the cold of the Arctic and Antarctic oceans and land around the Arctic. Plants and animals have a tougher time on the Antarctic continent, however, because there is both too much and too little water. The too much water is the ice that covers almost all of Antarctica. Although algae can scratch out a living in ice, no larger plants can. All forms of life need water in its liquid form. When the air temperature stays below 32°F (0°C) this is hard to come by. Hard, but not impossible to find.

Polar Talk

Plants that have vessels to transport internal fluids are called **vascular plants.** In addition to water-conducting tissues, most have true stems, roots, and leaves.

The Peninsula Is Antarctica's "Banana Belt"

By the standards of the rest of the world, the Antarctic Peninsula is a polar desert, but those who work elsewhere on the continent joke about it being Antarctica's "Banana Belt." This term applies best to the Peninsula's west coast, where the average temperature climbs above freezing for 3 or 4 months of the year and winter temperatures rarely fall below 15°F (–10°C). Annual precipitation is around 14 to 20 inches (35 to 50 centimeters) a year, about the same as in Denver, Colorado.

The Peninsula and the islands off its West Coast have patches of tundra, which are not as hospitable to life as Arctic tundra. Still, various kinds of lichens, mosses, and green algae cover many ice-free areas. In fact, in some places where water accumulates, you'll find patches of moss 3 feet (1 meter) thick. Antarctica's two species of vascular plants, Antarctic pearlwort and Antarctic hair grass, live here.

True Land Animals Are Hard to See

The only animal life that lives off land plants on the Antarctic continent are small *invertebrates*. These include maybe 60 species of nematodes, which are worms that live in the ground; 50 species of mites; about half as many species of springtails; and a few other tiny creatures. The largest of these are springtails, which are about half the size of a grain of rice.

To survive in bitter cold you need shelter, which tiny animals find in places such as between grains of sand, or under rocks that are semi translucent, allowing some sunlight through. Some insects live in lichens. Like some polar ocean creatures, some of Antarctica's small land animals have antifreezes that keep them from turning into chunks of ice when temperatures fall.

Polar Talk

An **invertebrate** is an animal without a backbone or spine. Such animals have less well-developed brains than vertebrates, the animals with spines.

Despite Appearances, the Dry Valleys Aren't Dead

The first explorers in the Dry Valleys, at the beginning of the twentieth century, saw no indications of life. But as scientists have been discovering since the mid-1950s, the Dry Valleys aren't dead. The few creatures that live there are hanging on at the very edge of life on Earth, however. Growth can be extremely slow with scientists estimating that one species of lichen, *Buellia frigida*, is growing as little as half an inch (1 centimeter) in 1,000 years.

The richest Dry Valleys' life is in lakes that have a year-round lid of thick ice, but never freeze all the way down. Don't look for fish in these lakes; all life there is microbial. It consists of mats of various organisms including *cyanobacteria* and diatoms. Researchers who don a couple layers of long underwear, dry suits, and Scuba gear to dive into these lakes descend into a world of life forms that dominated Earth until about 600 million years ago, when more complicated kinds of life began to evolve.

Polar Talk

Bacteria that live in water and which have the ability to create plant matter with photosynthesis are **cyanobacteria**—sometimes called "blue-green algae." They are single cells, but often grow in large colonies. They first appeared on Earth more than 3.5 billion years ago.

The mats in the lakes are a source of food for the largest land animals in the Dry Valleys, nematodes only about 0.04 inch (1 millimeter) long. Researchers have found that material from the mats comes loose and floats up against the bottom of the ice. As ice on top sublimates directly into water vapor in the dry air, and water freezes to the bottom of the ice, the material works its way to the top of the ice, where winds can blow the biological material around the valley to feed nematodes. The little bit of water that melts from the glaciers during summer's five or six weeks gives the

Polar Talk

Anhydrobiosis is a state of apparent suspended animation that some invertebrates enter to survive dryness or other extreme stresses. Animals in this state appear to be dead for days, weeks, or even years until moisture returns.

nematodes what they need. Diana Wall of Colorado State University says that when the nematodes sense that the environment is drying up, they go into a state called *anhydrobiosis* or life without water. They start to change their shapes into little coils that look "like a Cheerio" and begin producing antifreeze. "Nematodes in anhydrobiosis survive for years, until the environment becomes more favorable. They do have the fountain of youth worked out, it just takes them several years to know it!"

Antarctica's Ice Isn't as Dead as It Seems

Scientists have realized for decades that some forms of life can live in ice on land as well as in the ocean. On land, algae that thrive in the cold add a red tinge to parts of some glaciers. Microbes have been found in the snow at the South Pole, frozen in the ice of one Dry Valleys lake, and even deep inside the Antarctic Ice Sheet. When found, they appear dead, but come to life when warmed. Scientists are especially interested in these microbes because they might offer clues about what kinds of life, if any, might be found on Mars or elsewhere in the solar system. We take a closer look at these hardy creatures in Chapter 24.

Antarctica's Birds Dominate the Shores

Anyone who visits Antarctica has a good chance of seeing a whale or two in the ocean if much time is spent on board a ship, or even on land looking over water. Visitors are more likely to see a few seals on the ice if much time is spent near the coast. Anyone who sails the Southern Ocean or spends even an hour or two on a Southern Ocean island or near the Antarctic Peninsula Coast is sure to see birds, probably lots of birds, which nest on the islands and around the continent. Penguins nest in all these areas, but flying birds are harder to find the farther south you go.

Albatrosses Are at Home Over the Ocean

About all that most of us know about the birds called albatrosses is that saying, "he has an albatross around his neck" means someone is in deep trouble. In Samuel Taylor

Coleridge's poem "The Rime of the Ancient Mariner," sailors hung the carcass of the albatross around the neck of a shipmate who had killed the bird. They did this after things started going wrong. The disasters included the ship becoming becalmed in the tropics where there was:

> Water, water, every where,
> And all the boards did shrink;
> Water, water, every where,
> Nor any drop to drink.

Coleridge's poem is based, in part, on the old sailor's legend that seeing an albatross means fog or a storm was on the way, but killing one would make things even worse.

More than a half dozen species of albatrosses nest on the islands of the Southern Ocean, but the most impressive by far is the wandering albatross with its 12-foot (3.5-meter) wingspan.

Even more impressive than the size of wandering albatrosses is the amount of ocean they cover while wandering with the winds. In recent years, researchers have attached radio transmitters that communicate with satellites, enabling them to follow the large albatrosses. The birds tracked have traveled an average of 185 miles (300 kilometers) a day, sometimes going as far as 620 miles (1,000 kilometers) in a day.

Giant Petrels Help Keep the Southern Ocean Clean

Those who like to ascribe human attributes to birds could look on wandering albatrosses as free spirits of the oceans, tied to no land. Giant petrels, on the other hand, could be see as nasty creatures that you'd never want to invite to dinner. A more scientific view, of course, is to see these two species of large, Southern Ocean birds as creatures that have evolved different ways to make a living.

Giant petrels, which are the largest of the many kinds of petrels, make their living, at least in part, as scavengers. They eat dead penguins, seals, whales, or anything else that happens to die and float to the surface or wash up on land. When seals give birth on the ice, giant petrels arrive to eat placentas and dead pups. They are sometimes called the "vultures of the Antarctic," which means, if you want to give birds human virtues, they are good guys that help keep the Southern Ocean and Antarctica's coast clean. Giant petrels don't always wait for something to die, but have been observed attacking penguins and eating their chicks.

Explorations

Mimi Wallace, an El Paso, Texas, teacher who worked with scientists at the Palmer Station in January 2000, describes (on her website) a visit to a giant petrel nesting area: "You can walk amongst the birds and observe them with their chicks. The birds have been habituated by some of the scientists who study them. If the birds are annoyed or feel threatened they have the ability to induce projectile vomit! It is called gak! They are incredibly accurate. [The researchers] say the stench is sickening and you will never be able to wash it out of your clothes. [The birds] use this ability to vomit on demand to protect themselves. This vomit is oily and when it hits another bird's feathers, the bird will die. The oil saturates the feathers and when the bird gets wet it is not able to stay warm."

Some See Skuas as Gangsters with Wings

The south polar skua ranks with the leopard seal (see Chapter 7) as the Antarctic creature that people like to denigrate. "They are noisy, dirty and just plain nasty birds. They are to standard seagulls what a New York pigeon is to a dove," says Ethan Dick in the glossary that's part of his Antarctic journals website. Skuas are about the only flying birds seen as far south as the U.S. McMurdo station. People who work there tell stories of how skuas steal food, and like to "dive-bomb" people. Because most people seem to like penguins, the fact that skuas eat penguin eggs and chicks made them repulsive to some people.

South polar skuas are one of the seven species of skuas found in both polar regions, and they often travel long distances. South polar skuas are the only skua species that nest on the Antarctic continent. Because there are no trees or bushes, they nest on the ground in sheltered, shallow depressions, after arriving on the Antarctic Coast sometime from late October and mid-December. Some skuas feed heavily on penguin eggs and chicks, which have been abandoned by their parents and, thus, left defenseless. Other skuas eat mostly fish and krill in the ocean.

As the days grow darker and colder in March, skuas begin heading north. Not much is known about exactly where most of them go, but they seem to spend a lot of time at sea. They have been seen as far north as Alaska and Greenland during the Northern Hemisphere summer. During the Antarctic summer, a few skuas explore inland and a few have been seen at the South Pole Station, which is around 800 miles from the coast and the only place in Antarctica a skua will find something to eat. No one knows whether these wandering skuas ever make it back to the coast.

Everyone Seems to Adore Penguins

In her book *The Moon by Whale Light*, Diane Ackerman nicely sums up our attraction to penguins: "… they stand up straight and walk upright like humans, so we see them as little humanoids—a convention of head waiters, ten thousand nuns, plump babies wearing snow suits."

One of the reasons that penguins are cute is that they waddle when they walk, they're awkwardness is adorable; not what we expect in a wild creature. Because they can't fly, penguins would never survive if they had to chase their food on land. They don't, of course. In recent years, divers have filmed penguins underwater, and anyone who watches these films sees how penguins survive—they "fly" through the water with the grace of hawks in the air. Their streamlined bodies and movements make them fish-catching machines.

Penguins are strictly Southern Hemisphere birds—with one minor exception—that aren't confined to the ice. The exception is the Galapagos Penguin that lives on the Galapagos Islands, which straddle the equator in the Pacific. A few of these birds nest just north of the equator.

The other 16 species of penguins live on the shores of Africa, South America, Australia, New Zealand, and the islands of the Southern Ocean as well as the Antarctic continent. Of these, chinstrap and gentoo penguins live only on the Antarctic Peninsula, leaving Adelie and emperor penguins as the ones with colonies that circle the Antarctic Coast.

Adelie Penguins Are the Pioneers of Penguins

Adelie penguins live farther south than any other birds, and have been moving south in Antarctica since the ice started melting at the end of the last ice age. They nest on rocky coasts as close as possible to open water, but early in the season, before the sea ice has broken up, they might have to waddle 30 miles (50 kilometers) or farther across the ice to reach the water. Each summer they return to the same nesting place and mate, if they can. They usually lay two eggs and the males and females take turns incubating them, spending maybe 10 to 20 days on the nest without eating and then about the same at sea feeding. Chicks hatch in December and the mates take turns guarding the chicks and feeding at sea, swapping each couple days. When the adults come back to the nest they regurgitate food for their chicks. By the end of summer in February, the chicks are ready to swim north with the adult birds.

Emperor Penguins Are Real Tough Guys

Any time you start feeling down because life is hard, give a thought to how emperor penguins live. In May, when Antarctica is falling into winter with each day's high temperature likely to be below 0°F (–18°C), a female lays one egg, which her mate puts on the top of his feet, where the egg is covered by a patch of skin on his lower abdomen, which has a fold that covers the egg. The females head off to sea to spend the next three months or so—the depth of winter—feeding in the ocean while the males stand in huge groups, shuffling around to change positions, giving each bird time in the middle of the group, and time on the edge catching the bitter wind, with a chill as low as –75° F (–60°C).

An emperor penguin.

(Peter West, National Science Foundation)

Emperors' huddling makes them the only species of penguin that doesn't defend its territory. The tight huddling of emperor penguins generates enough warmth to cut each bird's heat loss by half. This means they are slower to use up their reserve of fat, since they aren't expending as much energy keeping warm. Females return in mid-July, finding their mates by recognizing their calls. The male transfers the egg to the female, without allowing it to fall on the snow, and then heads for the ocean to begin feeding for the first time in almost four months. Females returning from the ocean and males going to it, might travel more than 100 miles (160 kilometers) over the ice.

Evolving their unique nesting strategy gives emperors two advantages. No skuas or leopard seals are around when the birds are incubating eggs and soon after the chicks hatch. Also, the chicks have around five months of summer to grow large enough to move out on their own.

As you can imagine, emperor penguins are well insulated with four layers of feathers that are almost like scales, which even strong winds can't fluff up. Their very small bills and flippers mean little heat is lost through them. In addition, the arteries and veins going to and from the feet, wings, and bill are close together, which means blood is cooled on its way to the extremities and warmed on the way back to the heart.

The Least You Need to Know

- The Antarctic Peninsula is the continent's warmest area with the most land life.
- Plants and animals survive by sheltering among rocks and the soil, where tiny amounts of water are available for a small part of the year.
- Life in Antarctica's Dry Valleys survives on the very edge of existence.
- Bacteria have been found living on and deep within Antarctica's ice.
- Penguins and flying birds nest on Southern Ocean islands and around the Antarctic continent, with flying birds becoming less common the farther south you go.

Part 3

Going to the Ends of the Earth

Even if no one lived there or had never explored the Arctic and Antarctic, these regions would be fascinating places. Human life and exploration add another dimension to our interest. The people who later became the Eskimos and Inuit moved to the Arctic Ocean Coast around 5,000 years ago and developed ways to live in a frigid climate. Antarctica, on the other hand, is the world's only continent that has never had an indigenous people; in fact, no human even saw Antarctica until the early nineteenth century.

Chapter 10

The First People Move into the Arctic

In This Chapter

- People enter Siberia's Arctic 12,000 years ago
- Mongolians cross the land bridge to Alaska, sweeping south to become the first Native Americans
- Sami (Lapps) colonize Europe's Arctic Coast
- Tuniits pioneer the entire North American Arctic about 5,000 years ago
- Inuits take over Arctic lands about 1,000 years ago

To survive in the Arctic, people need ways to handle the cold, which folks first managed in Siberia. We see how the decreasing sea level during the last Ice Age created a land bridge that allowed humans to move into the Americas from Asia. These first arrivals moved south to become the people Europeans came to call "Indians." Starting around 5,000 years ago, people who had learned to live on the tundra and ice settled the Arctic from Alaska east to Greenland. More than 2,000 years ago, Inuits in Alaska learned to hunt bowhead whales, and spread to Greenland, replacing the earlier people. But climate cooling beginning around 700 years ago sent them into decline.

People Begin to Move into the Cold

By 50,000 to 35,000 years ago, during the last ice age, modern human beings, *Homo sapiens*, had developed language, stone tools, cooperative hunting skills, the use of fire, and warm fur clothes. They were ready to venture into cold country. Gradually, small bands of hunters followed game and settled in Europe, some nestling their homes almost against the ice sheet covering much of Scandinavia. When the ice retreated, humans occupied the areas it had covered, eventually reaching eastern Siberia from both Europe and Asia.

Around 35,000 years ago, during the ice ages, ice sheets were growing, drawing water from the oceans, which bared land like a drying puddle exposes mud clumps. The emerging land formed land bridges across areas that today are under water. From around 35,000 years ago to about 10,000 years ago—when melting ice sheets covered the bridge—waves of Mongolian peoples followed mammoth and bison herds to North America. The animals headed south looking for richer fodder, with the humans following them. These hunters became the first Americans, the ones we call "Indians."

Polar Talk

Homo sapiens are the modern species of human beings, the only surviving type.

People come to America.

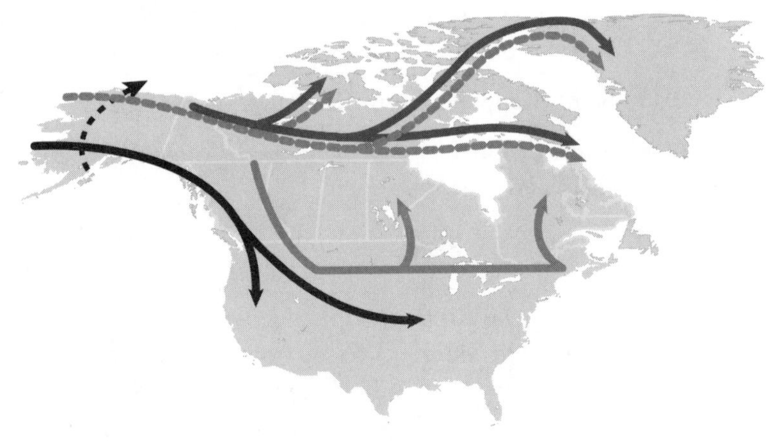

35,000 to 10,000 years ago: Native American ancestors.

10,000 to 5,000 years ago: Native Americans go north to tree line.

5,000 to 4,000 years ago: Tuniits go east.

3,000 to 2,000 years ago: South Bering Sea and North Pacific peoples meld into Inuits.

1,000 years ago: Inuits conquer fading Tuniits

In Europe, the Sami Settle Lapland

The *Lapps* (or *Sami* as they call themselves) are short, broad-headed folk with dark eyes and dark hair who moved into the northern forests of Norway, Sweden, Finland, and eastern Russia, known as Lapland, by 8,000 years ago. They may have come across the Ural Mountains or followed retreating ice and caribou north.

Earliest accounts say that 2,000 years ago, the women and men wore skin clothes sewn with sinews and fished and hunted caribou, bear, fox, wolverine, and beaver. Hunters speeding on skis, which they invented, could overtake animals. By 1,000 years ago, as wild game dwindled, Sami tamed caribou, which they used to pull sleds and for meat, milk, and skins.

Polar Talk

Sami (Lapps) are nomadic herders who inhabit Lapland in the northern Scandinavian Peninsula, including parts of Norway, Sweden, Finland, and Russia.

Arctic Dictates How People Make a Living, Live

The nature of the Arctic dictated how people without today's technology lived. In the past, they all fished and hunted sea mammals and land animals, following the animals made them a people on the move. They evolved similar technology: tailored skin and fur clothing, houses built partly underground, skin-covered tents, and skin- or bark-covered boats.

On the tundra, they used sleds and in the taiga: toboggans, skis, and snowshoes. Dogs carried loads and pulled sleds in both hemispheres. They worked stone similarly. They all used metals before the Europeans came. They even shared the same loose, casual social groups necessitated by their constant moves. All believed in a spirit world of life and death importance.

Explorations

The word *Eskimo* is commonly used for native people in Arctic North America and Greenland. Today, Canadian and Greenland natives consider it derogatory and use *Inuit* instead. Alaska's Arctic native people don't find the word offensive and use it more or less interchangeably with *Inupiat* to refer to themselves. *Eskimo* is considered derogatory because it's believed to come from an Indian word meaning "eater of raw meat." But the *American Heritage Dictionary of the English Language* says many linguists speculate *Eskimo* actually derives from an Eastern Canadian Indian word referring to a particular way of lacing snowshoes. In this book, where possible, we use the term that the people we are writing about prefer.

Tuniits Were First Year-Round Arctic Residents

Around 5,000 years ago, long after the land bridge had disappeared, Siberians crossed the Bering Sea by boat. Some reached northern Alaska and headed into the Yukon, becoming the first Eskimos, called Tuniits, and eventually pioneering the High Arctic from Canada to northern Greenland. They were the first people to actually live all year in the American Arctic.

CAUTION

Crevasse Caution

We are using *Tuniits* as a general term for several early Eskimo cultures, which archaeologists have identified. The many names can become confusing. Also, *Tuniits* has an Eskimo ring to it that some of the other terms, such as "Dorset culture" (one of the major ones) do not. If you haven't studied Arctic anthropology, *Dorset* might make you think of people living in thatched-roof cottages in England.

By 4,000 years ago, Tuniits had evenly settled the Arctic from Siberia east to Greenland. In some ways, it was easy. The climate was warmer than now. Along the coast, they found great piles of driftwood for fires. They shot musk oxen and seals.

The Right Clothing Makes All the Difference

An Eskimo, dressed for winter travel, looked almost as wide as high in his bulky double parkas made of caribou skin. He wore the inner one with the fur facing his body and the outer one with the hair facing away. Most inland folk wore parkas slit up the side to just above the waist. The front hung to mid thigh but the back reached to the back of his knees: an insulated seat cushion. His pants, made also of caribou skin, were big, floppy, and came to just above the knee. Underneath he wore leggings from ankle to thigh and a long stocking of soft skin. His feet were clad in caribou slippers, low-cut and soft, covered by two pairs of caribou boots, the first with hair side in and the second with hair side out. Hair covered the sole, too, so that layers of hollow caribou hair insulated his feet against the frigid snow. The Eskimo caribou clothing is lighter and warmer than anything else yet devised

When they decided to stay a while, they built a fireplace patterned on an ancient Siberian model. Living in oval tents, summer and winter, they divided the extended family tent in two with a central hearth made of stone, which people from both sides could reach. To boil food, they dropped hot stones into skin bags.

As time went on, neolithic geniuses devised technological improvements. Shooting an arrow or throwing a spear at a seal only to have it disappear with the imbedded weapon was catastrophic. You could starve that way. Some innovative hunter created the first harpoon. He flung it at a seal; the head broke away as the point pierced the animal. He held a line attached to the harpoon head. When the seal submerged below the ice to escape, the crafty hunter reeled it in.

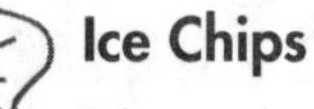

Ice Chips

Eskimos ate raw seal blubber all they could. They also ate great quantities of raw meat and fish. After all, they practically subsisted on meat and fish alone, and by eating as much raw stuff as possible they didn't kill the vitamins their bodies needed.

Climate Cooling Forced Innovations

The climate cooled about 3,500 years ago to present conditions, forcing the Tuniits to improve shelter and heating. They buried part of the home and built sod walls above the ground. Some savant invented the seal-oil lamp, a bowl containing burning oil that radiated a warm glow throughout the shelter. The lamp made an open hearth obsolete, because one could also cook over the lamp.

One thing remained: an ice dwelling easily built by a skilled snow cutter who could cut the right-size pieces and fit the tricky top blocks together. This shelter was reasonably warm (even without a fire), and ideal for a family that needed to make frequent winter moves as they chased seals. The Tuniit culture and its forerunners lasted from 5,000 to 1,000 years ago as the sole occupants of most of the Arctic. Their civilization died, probably after repeated failures in the caribou or seal hunts, about the time Norsemen and Inuits penetrated their land.

Crevasse Caution

Eskimos didn't push Grandma out on the ice floe to freeze when she was too old to sew. Eskimos value elders for their wisdom, knowledge of taboos and customs, and ability to tell a good story. It is true, however, when a family was literally starving, an elder might decide to wander out in a storm so the others could live. But all believed her soul would return when the next baby was born. She hadn't really died—just moved on.

Inuits Invent Dogsleds, Big Boats, and Wealth

When Siberians came to the New World about 5,000 years ago, some kept traveling—the Tuniits—to pioneer the Arctic. Others—the Inuits—stuck around the Alaskan coast, fought with each other, and warred and traded with others. Out of this turmoil emerged a superior Arctic technology. From the bones, teeth, and skins of animals, Inuits fashioned a better way of life.

Some of their inventions included large walrus skin boats called *umiaks*, dogsleds, whalebone lances and harpoon heads, *compound bows*, *crampons* for traction on ice, barbed arrowheads, and sharper knives made of slate.

These and other innovations opened the Arctic to the Inuit. Several men could glide their stable umiak onto a bowhead for the kill and they and their families would eat well for the whole winter. At day's end, the family could enter a warm pit house, loll on stone platforms covered with cushy furs, and toast their toes against a glowing soapstone lamp.

Polar Talk

A **umiak** is a large Eskimo boat made of skins stretched on a wooden or bone frame and is propelled by paddling. A **compound bow** uses a system of cables and pulleys to make the bow easier to draw. **Crampons** are spiked plates that fit onto shoes to prevent slipping on ice.

Having invented the tools, Inuit hunters went after bowhead whales. About 1,000 years ago, a warming climate opened leads in the ice of the Beaufort Sea and Amundsen Gulf. Bowhead whales migrated east through the leads, and the Inuits tailed their treasure from a home base in northern Alaska.

Whole families piled in umiaks, took off after their livelihood, and settled the Arctic along the way. They reached Labrador about 750 years ago. Their rich and comfortable culture dominated for 500 years and then declined, with the weather.

Polar Talk

The **Little Ice Age** was the general cooling of Europe and other parts of the Northern Hemisphere that began about 700 years ago and lasted until about 150 years ago. The coldest period began about 500 years ago.

The Little Ice Age Hits the Inuits

Beginning around 700 years ago, the Northern Hemisphere began cooling, the Arctic with it, during the period we call the *Little Ice Age*. Ice advanced and closed leads. By this time, European whalers began to hunt whales in the western Atlantic, leaving fewer to migrate west into Inuit waters. The Inuits—no longer able to hunt the economic foundation of their society—changed their lives.

Many folks in the High Arctic died trying to adapt. Survivors abandoned the area and moved east into northwestern Greenland where they eked a living hunting seals and gathering bird eggs.

Explorations

Eskimos did lend their wives to strangers, but for a good reason: to stay alive. With no government to keep the peace, a man's avenging kin was his safety net. The threat of reprisal would keep a potential criminal in line. However, an Eskimo who left his homeland to trade or hunt was among strangers who might kill to get his dog team. He made a home-away-from-home safety net by taking his wife. In a new place, the man arranged a wife exchange with a local resident. That made him kin to all his wife-exchange husband's family. That family was bound to avenge any injury done to him.

In the central Arctic, summer ice increased so much that Inuits gave up hunting bowhead whales and turned to the interior for food. Fishing inland waters and hunting caribou, they scrounged only small stores for the winter. When these provisions ran out, they followed the ringed seal.

Firm ice—the ringed seal's favorite habitat—expanded and lasted longer into summer as the climate cooled. Inuits who switched to *breathing-hole hunting* did well; others did not. Reluctantly, they abandoned their fine winter homes and lived in snow houses on the ice. They could build a snow house in about an hour, which meant moving was no great chore. When they ran out of seals in one place, they moved to another.

Polar Talk

A seal makes many holes to breath at, and a hunter using a harpoon—a spear with a barbed head—waiting at one to strike if the seal surfaces is **breathing-hole hunting**.

Inuits dropped whale-hunting technologies. Culture, art, and mythology shrank. Most Inuits were living at a subsistence level when European explorers arrived.

Whalers and explorers brought goods to trade: guns, cloth, metal, tools, and utensils. They also brought a different way of life and thinking: clocks, musical instruments, and dances. Finally, they brought unintended scourges: disease, tobacco, and alcohol. In Chapter 15, we will see how the Inuit live today, keeping parts of their old culture alive, but very much a part of today's world.

The Least You Need to Know

- The first people entered the Arctic in northern Siberia about 12,000 years ago.
- Mongolians crossed a land bridge to the American Arctic in waves that pushed south, becoming the Indians.
- Siberians crossed the Bering Strait by boat 5,000 years ago and became the New World's Eskimos.
- The first people to spread across the Arctic from Alaska to Greenland were the Tuniits.
- The Inuits learned to hunt whales and spread from Alaska to Canada and Greenland.
- The Little Ice Age, beginning 700 to 500 years ago, made whale hunting difficult; Inuit society went downhill.

Chapter 11

Europeans Head to the Ends of the Earth

In This Chapter

- The dream of a Northwest Passage between Europe and Asia sends explorers into the Arctic
- Nineteenth-century explorers find the great southern continent the Greeks had speculated about
- Loss of the 1845 Franklin expedition in the Arctic stuns Victorian England and spurs more exploration
- Important scientific work, such as finding the North Magnetic Pole, is accomplished
- The Northeast and Northwest Passages are navigated, but found to have no commercial value

Beginning with the ancient Greeks, educated people theorized about the northern and southern ends of the earth. But because no one traveled to the Arctic or Antarctic, they were free to speculate without any inconvenient facts getting in the way of their theories. From the fifteenth through most of the nineteenth centuries, Europeans—mostly English—explorers often suffered and died because they refused to meet the Arctic on its terms, before they began to learn how to live in the Arctic from the Inuits.

A Dream Drives Men Deeper and Deeper into the Ice

During the late 1400s, explorers from Portugal and Spain found routes around the tip of Africa to Asia, and "discovered" the New World, which redefined Europe's economy and politics by opening up new trade routes to Asia and building a Spanish Empire in the Western Hemisphere. The Northern European commercial powers—the English, the French, and the Dutch—needed a route to the Orient that didn't cross seas controlled by the Spanish, such as a Northwest Passage. During this time, an expanding Russia searched for a Northeast Passage.

Polar Talk

The **Northwest Passage** is an Arctic Ocean ship route between the Atlantic and Pacific oceans along the northern coasts of Canada and Alaska. The **Northeast Passage** is the ship route from the Atlantic to the Pacific Ocean along Scandinavia's ice-free Arctic Coast and Russia's Arctic Coast.

Go Around, Not Through America

As the English, French, and Dutch explored what is now the United States and the bordering provinces of Canada during the sixteenth century, each large river or bay raised hopes that the passage to the Orient had been found. But by late in the sixteenth century, it was obvious that a Northwest Passage, if it existed, would be across the Arctic. Thus began 300 years of exploration that left the Arctic littered with bodies and names of explorers.

England Leads the Search for a Passage

During these 300 years, most of the expeditions looking for a Northwest Passage were English. The story of Henry Hudson is an example. In 1610, a group of English merchants hired him to search in the north. He sailed through what is now Hudson Strait into what is now Hudson Bay, where the ship was trapped by ice for a winter of cold, disease, and hunger. In the spring of 1611, the crew mutinied rather than continue exploring. They set Hudson, his son, John, and seven sailors adrift never to be seen again.

Czar Peter the Great of Russia selected Vitus Bering, a Dane, to find a Northeast Passage. On journeys beginning at Kamchatka on Russian's Pacific Coast in 1728, he sailed into the strait between Russia and Alaska that's named after him. From 1733 until 1741, his Russian expeditions mapped Siberia's Arctic Coast, before dying in 1741 on the island now named for him.

By the end of the eighteenth century, explorers were filling in more and more of the map of the Arctic, but the Northwest and Northeast Passages remained elusive. From the start of the American Revolution in 1776 through the end of the Napoleonic Wars in 1815, the British Navy was too busy fighting to go exploring.

After 1815, with Napoleon safely tucked away in exile on St. Helena, and the quarrels with the Americans (the Revolution and War of 1812) settled, the British navy ruled the waves with not a whole lot to do. Arctic exploration was one of the things Sir John Barrow, the second secretary of the admiralty from 1807 to 1848, found to keep the navy busy. Barrow enjoyed the bureaucratic power and prestige needed to launch navy Arctic exploration.

In return, explorers attached his name to places they discovered. Thus, we have Cape Barrow and Barrow Strait in the Canadian Arctic, and Point Barrow and the town of Barrow in Alaska.

Hazards and Delusions Bedevil Explorers

Stories of polar exploration fascinate us because of the hazards and hardships involved. The hazards go beyond the cold, but they begin there. Anyone who ventures into the cold, whether he's a child going outside to play in the winter, or an explorer heading for the North Pole has to watch out for *frostbite* and *hypothermia*.

Polar Talk

The freezing of the skin or tissues under the skin is **frostbite.** If the tissue isn't warmed soon enough, it can die. **Hypothermia** is a decrease in the core body temperature to a level at which normal muscular and cerebral functions are impaired. It occurs when the body loses heat faster than it is replaced. Symptoms begin when the core body temperature drops below 95°F (35°C).

A mother can bring the child indoors when she sees the tip of his nose turning white, a first sign of frostbite, or when he begins shivering, a first sign of hypothermia. Polar explorers usually didn't have the option of a warm place to escape to. Over the centuries, many lost fingers and toes, and some feet and hands to frostbite. An uncounted number died of hypothermia. By the time Europeans began exploring the Arctic in the sixteenth century, Eskimos had developed the technologies for staying warm on the tundra and the ice. But until well into the nineteenth century, most European explorers, especially the large expeditions sent by the British navy, showed little interest in adapting Eskimos' ways of dressing, living, and traveling in the Arctic.

Even Getting a Drink of Water Can Be Hard

You might think that getting a drink of water would be little problem in a world of snow and ice. But putting cold ice or snow in your mouth can cause frostbite, and uses heat the body can't spare. Fuel needed to heat shelters, cook food, and melt ice for drinking water added to explorers' loads. Even worse, few Europeans used dogs to pull their sleds until late in the nineteenth century. Europeans exhausted themselves pulling sleds.

CAUTION

Crevasse Caution

If you ever get a chance to walk around on a glacier or ice sheet—say on a tourist excursion—heed the lessons learned by explorers: Don't let reflected sunlight burn the insides of your nostrils or eyelids. Today, thankfully you aren't likely to sunburn the roof of your mouth as you pant with the exertion of pulling a heavy sled across jumbled ice, as many nineteenth-century explorers did.

The ever-moving Arctic sea ice was the greatest obstacle to European explorers. A channel that was open last summer can be ice-choked this year. As currents and winds move the ice, colliding floes build pressure ridges. Imagine what happens to a ship caught between two floes. Many ships ended up on the floor of the Arctic Ocean, crushed.

Open Polar Sea Delusion Was Slow to Die

Until at least the 1870s, many argued that ships would find open water once they pushed through the sea ice that had stopped other explorers. The idea of the open polar sea may have come from the ancient Greeks, who thought that an ocean flowed around the world, and later map makers continued showing it. In the nineteenth century, the open polar sea was scientifically respectable, although many who knew the Arctic disagreed. August Petermann, a German geographer and one of the most important map makers of his time, theorized that the warm water the Gulf Stream brings north flowed between Greenland and Siberia, opening up the ocean. Matthew Fontaine Maury, who is considered the founder of physical oceanography, devoted a chapter of his 1855 book, *Physical Geography of the Sea,* to the open polar sea idea. He, too, wondered where the heat brought north by the Gulf Steam went.

Meanwhile, at the Other End of the Earth …

Our word *Antarctica* comes from the Greek *Anti-Arkikos*, or the opposite of the Arctic. Greek thinkers believed Earth needed a large southern Continent to keep in balance. With nothing more than this to go on, some early map makers showed such a continent extending almost as far north as the tip of South America.

By 1644, explorers had shown a southern continent was not connected to South America, Australia, or Africa. From 1772 to 1775 the Great British explorer Captain James Cook sailed completely around Antarctica, and his ships were the first to cross the Antarctic Circle, reaching 71 degrees south before being blocked by ice. During his voyages in the Southern Ocean, Cook never saw Antarctica. He believed ice would keep anyone from ever exploring the continent, if it existed.

Cook reported large numbers of seals in the Southern Ocean, which set off a "seal rush." As they killed off all of the seals on the sub-Antarctic islands, sealers sailed south, looking for new hunting grounds. What they actually discovered is hard to determine because a sealer who found an island covered with seals wasn't likely to tell others where it was.

For all we know, some now-unknown seal hunter was the first human being to see Antarctica. For a long time, U.S. maps showed the "Palmer Peninsula" because of the claim that Nathaniel Palmer, a young American seal captain, was the first to see the Antarctic Peninsula, and thus a part of the Continent, on November 17, 1820. As far as we know, Palmer himself never made this claim.

Earlier in 1820, a Russian expedition led by Thaddeus von Bellingshausen, and a British expedition led by Edward Bransfield, were exploring the ocean near the Peninsula, and people from one of these expeditions, not Palmer, might have been the first to actually see Antarctica. At the time, no one seemed to care. Both the British and Russian expeditions returned home to discover that no one was very interested in their reports.

Magnetic Poles Attract Explorers to North and South

Like many European ships used to search for a Northwest Passage, the steamship *Victory*, commanded by John Ross, became trapped in the ice in 1829, but unlike many of the others, Ross's men survived, not merely 1, but 4 winters in the Arctic, with only 3 of the original 19 men dying. Ross and his men established friendly relations with Inuit in the area, learning how to live in the Arctic.

Polar Talk

The points in the Arctic and Antarctic toward which the needles of magnetic compasses point are the **magnetic poles.** They are not at the geographic poles, which is why compass needles do not point due north and south from most places on Earth.

Ross's nephew, James Clark Ross, was second in command and he realized from the way the ship's compass acted that the North *Magnetic Pole* might be nearby. Expeditions at the time normally carried magnetic observation instruments, and with these Ross found the North Magnetic Pole at latitude 70 degrees, 5 minutes north, and longitude 96 degrees, 47 minutes west. The public and the government acclaimed the expedition's successes, and the younger Ross was selected to lead an expedition to find the South Magnetic Pole.

For his journey south, James Ross had two ships, the *Erebus* and *Terror*, which had been strengthened to withstand battering by ice. Soon after crossing the Antarctic Circle on New Year's Day 1841, they reached the pack ice that had stopped all explorers since Cook from going any farther south. The ice-strengthened ships enabled Ross to reach relatively clear water in what is now the Ross Sea. The expedition sailed south to McMurdo Sound (which Ross named after the first lieutenant on his ship, Archibald McMurdo). Ross failed to reach the South Magnetic Pole because it's inland, but his instruments showed that it was somewhere in the neighborhood, generally speaking. By reaching McMurdo Sound, Ross pointed the way to the South Pole for those who followed a half-century later. Ross Island is one of the closest locations to the South Pole—about 900 miles—reachable by ship.

Back to the Arctic, and a Disaster

After spending three years exploring the Antarctic and the Southern Ocean, James Ross with the *Erebus* and *Terror* returned to England in September 1843, all but ending Antarctic exploration for the next half-century.

In terms of the number of ships and men involved, however, polar exploration reached a peak in the last half of the nineteenth century that wasn't to be matched until after World War II. The rush to the Arctic began when Sir John Barrow, the second secretary of the British Admiralty, appointed Sir John Franklin, a polar veteran, to take the *Erebus* and *Terror*, with 129 men to find the Northwest Passage. Failure wasn't contemplated.

When three years passed with no word from the expedition, concern begin to grow, and the British navy sent out three expeditions, including one led by James Clark Ross, to search for Franklin, with no success. They were the first of more than 20 expeditions to search for any signs of Franklin's men and ships.

In August 1850, searchers found supplies and scraps of clothing from the *Erebus* or *Terror* and signs of a camp on Devon Island. Later that month, they found three graves with wooden headstones marked "1846" on Beechy Island. This suggested that the ships had been trapped in the ice nearby the first winter, but offered no clues as to what went wrong. Over the next few years, searchers encountered Inuit who told stories of groups of white men dying on the ice.

Finally, in 1859, searchers found a note in a cairn—a pile of rocks—saying that Franklin had died in June 1847, and that on April 22, 1848, the 105 survivors had abandoned the ships, which had been stuck in the ice since September 12, 1846. They were apparently heading south with the hope of reaching lands were game could be found to keep them alive as they tried to reach a Hudson's Bay Company outpost around 1,200 miles away.

Explorations

The mystery of what happened to the Franklin expedition still fascinates, with new books coming out regularly and the search still going on for the *Erebus* and *Terror* under waters in the Canadian Arctic. In the 1990s, analyses of the bones of three Franklin crewmen, whose bodies had been exhumed in the 1980s, corroborated the hypothesis that lead poisoning had killed them. It would have been caused by poorly applied beads of lead solder on the cans of food the expedition carried.

Government Interest in Arctic Exploration Wanes

While all the expeditions searching for any sign of the Franklin party filled in major gaps in the map of the Arctic, they also showed that although it probably would be possible to sail through a Northwest Passage, it would be of little or no commercial value. Governments stopped sending large expeditions into the Arctic. From late in the nineteenth century until after World War II, Arctic and Antarctic explorers—including the big names—had to raise private funds to pay most of their expenses. Although some governments gave some support, much of the funding was private, often from companies that wanted to advertise that a polar explorer used their shirts, socks, or soup.

Two Scandinavians Made the Passages Look Easy

After three centuries of striving and suffering and failing to find the Northeast and Northwest Passages, success seemed somewhat anticlimactic, and anyone but dedicated students of Arctic exploration would be hard pressed to tell you who did it. The Swede and Norwegian who finally made it were such skillful polar explorers that they made it look easy.

Nils Adolf Erik Nordenskiold, a Swede, and his crew left Norway in their ship, the *Vega*, in 1878 and went around Cape Chelyuskin, the northernmost point of Siberia, but were caught in the ice west of the entrance to the Bering Strait. They spent the winter with the Chukchi native people of the area before sailing through the Strait to reach Yokohama, Japan, and then returned through the Suez Canal to Stockholm, becoming the first to circumnavigate the entire Eurasian landmass.

Roald Amundsen and six crewmen left Oslo, Norway, on June 16, 1903, in the small fishing boat, the *Gjoa*. They sailed into the Canadian Arctic and after a couple close calls, an engine room fire and almost grounding on a reef, reached a bay, which Amundsen named Gjoahaven, or a haven for *Gjoa*.

Ice Chips

When Roald Amundsen reached the location near where Ross had found the North Magnetic Pole in 1831, he set up a magnetic observatory and found that the pole had moved to the north. Amundsen spent almost 2 years taking observations, which led scientists who examined them to conclude that the pole had moved about 31 miles (50 kilometers). This was the first evidence that Earth's core is not solid.

Amundsen spent the two years in Gjoahaven taking magnetic observations that showed the North Magnetic Pole had moved since Ross found it, and also learning Arctic survival and travel techniques from the Inuit, skills that would help in his later explorations. They completed the Northwest Passage trip in 1906.

The Least You Need to Know

- Europeans spent three centuries searching for Northwest and Northeast Passages across the Arctic.
- The belief in an "open polar sea" around the North Pole lasted for centuries.
- During his 1772 through 1775 expedition, British captain James Cook became the first to cross the Antarctic Circle.
- Who actually first saw Antarctica isn't clear.
- A British Arctic expedition led by John Ross discovered the North Magnetic Pole in 1831.
- Roald Amundsen, a Norwegian, and his crew navigated the Northwest Passage between 1903 and 1906.

Chapter 12

A Century Begins with Heroic Age of Exploration

In This Chapter

- Ernest Shackleton gets within 94 miles (151 kilometers) of the South Pole
- Frederick Cook and Robert Peary, on separate expeditions, claim to reach the North Pole
- Roald Amundsen and four other Norwegians reach the South Pole
- In 1912, Robert Scott and four others reach the South Pole, but all die on the way back
- In 1915, Ernest Shackleton, who had planned to cross Antarctica, is forced to abandon his ship when ice crushes it

The twentieth century began with the polar regions as the only large parts of the earth that had never felt human footsteps. A burst of exploration in the years before World War I started in 1914, and filled the last big gaps in knowledge of these areas. The period is known as the Heroic Age of Antarctic exploration and the term is sometimes applied to the Arctic. This period produced larger-than-life men whose virtues and shortcomings are still hotly debated.

The Time Is Ripe to Reach the Poles

When the Sixth International Geographical Congress in London in July 1895 adopted a resolution calling Antarctic exploration "the greatest piece of geographical exploration still to be undertaken," it reflected a growing opinion that the time was ripe to push on to the ends of the earth. At this time, no one had been within 400 miles (643 kilometers) of the North Pole, and the South Pole was more than 900 miles (1,450 kilometers) from the nearest human footprints.

Attendees at the 1895 London conference had no way of knowing that three months before, on April 8, 1895, Fridtjof Nansen and Hjalmar Johansen had used dogsleds to get within about 250 miles (402 kilometers) of the North Pole. Nansen and Johansen had left Nansen's ship the *Fram*, which was frozen in the ice, drifting with the Arctic Ocean's currents. They didn't reach a place where they could communicate their news to the rest of the world until August 1896.

Nansen's and Johansen's journey is a rip-roaring adventure, including several close calls. It's also a story of men who stayed in the Arctic for more than a year, living on what they and their dogs could drag on sleds, and the seals and polar bears they killed. They had learned from the Inuit, who had been doing this for centuries.

Ice Chips

Polar explorers weren't able to announce their discoveries or radio for help all through the Heroic Age, even though radio (called wireless then) was being developed. Douglas Mawson's 1913 Australian expedition to Antarctica was the first to experiment with radio in Antarctica.

As the century ended, others such as the American Robert Peary and his African American associate Matthew Henson were learning from the Inuit. Peary referred to Henson as his "servant," but Henson was the one who learned the Inuit's language and gained their trust. Peary and Henson, unlike many earlier explorers, took Inuits on their journeys.

The British Send a Major Expedition South

When Robert Falcon Scott, with about 50 men, anchored his ship, the *Discovery*, in Winter Quarters Bay on Ross Island in January 1902, and erected a prefabricated building, he opened the exploration of the Antarctic continent. Scott, like almost everyone else on the trip, was new to polar exploration, and much of the story of the 1902–1904 expedition is of novices learning the ropes of Antarctic exploration—not always too well.

Almost all the expeditions that tried to reach the North or South poles during the Heroic Age used the same technique. A relatively large party would set out from the base camp, carrying as much food, fuel, and other supplies as they could. At regular intervals, depots of food and fuel would be established and part of the party would return to the base.

On the generally featureless ice, the depots would be marked with huge cairns made of snow and ice—the only materials at hand—and marked with colored flags that would stand out against the white background. Missing a depot and its vital food and fuel on the way back could be fatal.

Scott and His Men Head South into the Unknown

On November 2, 1902, Scott, Ernest Shackleton, and Edward Wilson (a physician) left the *Discovery* base on Ross Island headed for the South Pole. On November 15, the last of the supporting parties turned back, leaving Scott, Shackleton, and Wilson to push on toward the Pole. They, of course, had no idea what they would encounter. Was the ice flat, like the Ross Ice Shelf, all the way to the Pole? Or would they run into mountains too steep and icy to climb?

Today, when people in the most remote parts of the world easily call for help, it's hard for us to imagine the danger Scott, Shackleton, Wilson, and other Heroic Age polar explorers faced. If anything went wrong no one else would know until the party failed to return. In 1902, Scott, Shackleton, and Wilson nearly faced the dilemma of what to do when one of three becomes too ill to continue. By late December, it was becoming obvious that Scott had miscalculated how much food the men would need. All were suffering from sunburn and probably dehydration.

Illness Nearly Dooms the Three Explorers

As if this were not enough, Scott and Shackleton, to put it mildly, were incompatible. Scott was Royal Navy, Shackleton from the merchant marine. This alone implied much in terms of social differences in the extremely class-conscious England of the time, but Wilson managed to keep the peace.

On Christmas Eve 1902, Wilson saw that both Scott and Shackleton were coming down with scurvy, and on December 31, 1902, Scott decided to turn back before they reached the Transantarctic Mountains. The three men had reached latitude 82 degrees, 15 minutes south latitude. Scott and Wilson went on a little more than 2 miles farther, to 82:17 South, or about 532 miles (856 kilometers) from the Pole, leaving Shackleton to watch the dogs. When the last of the dogs died, the three had to pull the sleds themselves, but Shackleton became too ill to pull. At least he was able to walk much of the time. Scott, Shackleton, and Wilson barely made it back.

Shackleton Returns, Almost Makes South Pole

In January 1909, Shackleton, Jameson Adams, Eric Marshall, and Frank Wild got within 94 miles (151 kilometers) of the South Pole, when Shackleton decided they must turn back. "I thought you'd rather have a live donkey than a dead lion," Shackleton later told his wife, Emily. Still, Shackleton returned to a hero's welcome in England. Not only had he made a huge advance toward the Pole, but had mapped a route through the Transantarctic Mountains, up the Beardmore Glacier (and its treacherous crevasses), to the polar plateau, where the Pole sits.

Frederick Cook Claims to Be First to the North Pole

By 1909, if anyone seemed destined to reach the North Pole first it was the American Robert E. Peary. Since making his first trip to the Arctic in 1886, he had led four other expeditions to the Arctic, mostly to Greenland and the adjoining parts of Canada. He and his associate Matthew Henson had learned from the Inuit.

Yet when the world first heard that someone had reached the North Pole, it was Frederick Cook, an American physician. The newspapers told Cook's story of having left Etah, in northwest Greenland, on February 19, 1908, with 2 Inuit, 2 sleds, and 26 dogs. They crossed the ice to Ellesmere Island, and then headed north to the Pole, arriving on April 21, 1908. Cook didn't reach a place in Greenland where he could get word out to the world until May 1909. Cook was honored in Europe before returning to the United States where 100,000 people are reported to have cheered his arrival in New York City.

Meanwhile, on February 28, 1909, Robert Peary with 23 men, 133 dogs, and 19 sleds headed north from the northern coast of Ellesmere Island. After a week's delay while waiting for a large lead to freeze over, Peary and his party reached a point about 150 miles from the North Pole, where the last of the supporting party turned back. From here, Peary, Henson, and four Inuits headed north, where Peary says they reached the North Pole on April 6, 1909.

When Peary returned to Greenland, he heard of Cook's claim, and sent off telegrams saying he had reached the Pole. Not only did Peary say he had reached the North Pole, he said his party was the first there; that Cook had not made it to the North Pole. The dispute was bitter, with strong opinions on both sides, and it continues that way even today. Today, many of those who look at the issue doubt both claims because neither Cook nor Peary produced log books indicating the careful navigation needed to reach the North Pole over ice that was always moving.

Amundsen Makes It a Race to the South Pole

When the Cook and Peary claims hit newspaper front pages, Roald Amundsen had arranged with Fridtjof Nansen to use his ship, the *Fram*, for another drifting trip in the Arctic's ice. Amundsen also would surely attempt to reach the North Pole. The Cook and Peary claims changed Amundsen's plans. He turned his thoughts south, but he told only a very few associates. As far as the world was concerned, including Scott in England who was putting together his second Antarctic expedition, Amundsen was heading north.

Amundsen in the *Fram*, with 19 men and 97 carefully selected sled dogs from Greenland, left Christina (now Oslo), Norway on August 19, 1910, about 8 weeks after Scott's *Terra Nova* with 65 men had left Cardiff, Wales, for Antarctica. When the *Fram* stopped at the island of Madeira for supplies, Amundsen told the crew they were going to Antarctica to try to reach the South Pole. He released the news to the world and sent a telegram to Scott, which the British leader received when he arrived in Melbourne, Australia. The British were outraged. Antarctica was supposed to belong to them.

When the *Fram* arrived at the Bay of Whales—an indentation in the Ross Ice Shelf about 400 miles east of Scott's base on Ross Island—on January 14, 1911, Amundsen and his men set up their "Framheim" camp, and the men started hauling supplies inland and setting up depots. Amundsen with Sverre Hassel, Helmer Hanssen, Oscar Wisting, and Olav Bjaaland left Framheim on October 19, 1911, for the South Pole. They had some hairy moments, especially while crossing what they named the "Devil's Ballroom," a glacier cut by several deep crevasses. The Norwegians traveled only about five hours a day, spending the rest of the time sleeping, eating, or resting to conserve energy. No one wrote in his diary about being hungry; they were carrying plenty of food. As the loads became lighter they killed dogs to feed the remaining dogs. It wasn't a walk in the park; the men suffered minor frostbite and sunburn, but none was ever too ill to continue.

After the five Norwegians reached the Pole on December 15, they spent a full day verifying their navigation with Amundsen and three of the others, who were also qualified navigators, taking and recording their readings. Three men were also sent out for a few miles in different directions to ensure the party had really walked on the Pole if their navigations were in error. Before leaving, they set up a tent and left a letter to the King of Norway inside along with a note asking Scott to please deliver it.

Ice Chips

So we arrived and were able to plant our flag at the geographical South Pole. God be thanked.

—Roald Amundsen, December 15, 1911

Scott and His Men Show How Much Humans Can Endure

Unlike Amundsen, who made sure he purchased the best Greenland dogs for his expedition, Scott didn't trust dogs to pull sleds, although he brought some with him. He believed that Manchurian ponies would be best, since they did well in the Asian cold, which of course is not as cold as Antarctica, and the ponies failed to live up to his expectations. Scott and his men ended up pulling sleds loaded with supplies. This labor made them sweat, dehydrating them and also increasing their need for food. The theme of hunger runs through their diaries.

Scott and the four men with him arrived at the South Pole, frostbitten, hungry, and sick on January 18, 1912, to find the tent the Norwegians had left.

Ice Chips

Great God! This is an awful place, and terrible enough for us to have laboured to it without the reward of priority.

—Robert Falcon Scott, January 18, 1912

On the way back, Edgar Evans, a British navy petty officer, collapsed and died on February 17, 1912. A month later, Titus Oates, an army officer whose feet were frostbitten too badly for him to walk and pull a sled, said, according to Scott's diary, "I am just going outside and may be some time." He was never seen again. A week after that, a blizzard trapped Scott, Edward A. Wilson, and Henry R. Bowers in their tent only 11 miles from a large depot of food and fuel. As far as we know, the blizzard was blowing too hard for the three men to struggle the last 11 miles.

The rest of the expedition had waited all winter in their hut at Cape Evens on Ross Island. They knew Scott and the other four were dead; there was no way to survive on the ice with the supplies of food and fuel they had. Their search party found the tent with the bodies of Scott, Wilson, and Bowers and the diaries that told their stories on November 12, 1912.

The bodies of Scott, Wilson, and Bowers were left in the tent, with a huge snow cairn built over it, as their tomb. Since 1912, snow has buried it as the ice creeps inexorably toward the ocean 175 miles away. Charley Bentley, a University of Wisconsin glaciologist, estimated in 1999 that the ice cairn was then about 75 feet under the surface and about 30 miles closer to the ocean than in 1912. Bentley estimates the tomb could reach the coast about 275 years from now, probably to drift north over the Ross Sea deep inside a melting iceberg.

Ice Chips

[B]ut for my own sake I do not regret this journey, which has shown that Englishmen can endure hardships, help one another, and meet death with as great a fortitude as ever in the past … Had we lived, I should have had a tale to tell of the hardihood, endurance, and courage of my companions which would have stirred the heart of every Englishman.
—Robert Falcon Scott, from the diary found with his body

Shackleton Ends the Heroic Age with a Survival Tale

Even though Norwegian and British flags had been planted at the South Pole, Shackleton still wanted to win glory in Antarctica. His plan: cross the continent from one side to the other with the South Pole as a stop along the way. Shackleton and his party would sail on the *Endurance* to land on the then mostly unexplored Weddell Sea Coast, and head for the Pole. From the Pole, they would continue north along Scott's route to the Beardmore Glacier, where they would meet the expedition's Ross Sea party, which would lay supply depots from Ross Island to the glacier. World War I broke out as the *Endurance* was ready to leave England, but the government told them to go ahead to Antarctica.

Several books, movies, and television documentaries have told the *Endurance*'s story. Here's the outline: Ice closed around the ship in the Weddell Sea on January 18, 1915. In late October, the ice began crushing the ship, and on October 27, 1915, Shackleton and the 27 other men on board abandoned the ship with lifeboats and as much of their supplies as they could and set up camp on the ice before the *Endurance* sank. In late December, as the ice was breaking up, the party started toward the nearest land, pulling the lifeboats over the ice. In April 1916, all 28 men made it in the lifeboats to land, but it was remote Elephant Island north of the Antarctic Peninsula. On April 24, 1916, Shackleton and five others left in the 22-foot lifeboat *James Caird* for a whaling station on South Georgia Island. After 17 days crossing 800 miles of the stormy Southern Ocean, and in a feat of navigation that seems a miracle, the Caird reached South Georgia. A ship finally took the remaining 22 men off Elephant Island on August 30, 1916. All aboard the *Endurance* made it back home.

The Ross Sea Party Struggles to Do Its Duty

When the ship *Aurora* arrived in McMurdo Sound from Australia in January 1915 with Shackleton's Ross Sea party, those aboard had no way of knowing that the *Endurance* was trapped in the ice on the opposite side of the continent. They assumed that within a year Shackleton and his party would be counting on the depots the Ross Sea party were to set up.

The next two months turned into what seemed to be typical of British Antarctic expeditions: men struggling to haul loaded sleds across the ice, not enough food, frostbite and the beginnings of scurvy. Still, by the time the Ross Sea party returned to spend the winter in huts built by Scott on his 1902–1904 and 1910–1912 expeditions, they had left supplies as far south as 690 miles (1,110 kilometers) from the South Pole.

Wind, Ice Carry Away the Party's Ship, Supplies

By May, the *Aurora* was frozen into the ice at Cape Evens, secured to the shore with seven steel hawsers running from the stern to shore, two anchors at the bow, and another heavy cable from the middle of the ship to shore. Ten men were on shore in huts from Scott's earlier expeditions, with food and some supplies, but with much equipment such as extra clothing and stoves still on the ship. On May 6, 1915, a storm—winds might have hit 120 mph—snapped the cables holding the *Aurora* and pushed it along with the ice out to sea. Those ashore assumed it had sunk. Instead, over the next 10 months the *Aurora* drifted 1,100 miles (1,770 kilometers) north, locked into the ice. The 10 men left ashore spent the rest of the winter repairing sleds and stoves, left behind by Scott's expeditions, and making clothing out of tents and boots from extra sleeping bags. The area offered plenty of seals to be killed for food, and seal blubber for fuel.

With no way of knowing what had happened to the *Endurance*, the 10 men hauled food and other supplies for depots, with the one farthest south being about 230 miles from the South Pole. One man died during the depot-laying journeys, and two more disappeared—they probably fell through thin ice in May 1916. The *Aurora* finally broke free of the ice and made it to New Zealand. Ernest Shackleton himself was aboard when it sailed back to McMurdo Sound to rescue the seven survivors of the Ross Sea party in January 1917.

One Who Was There Sums Up Heroic Age Heroes

Apsley Cherry-Garrard was a near-sighted, young Englishman on Scott's 1910–1912 expedition and was with the party that discovered the bodies of Scott, Wilson, and Bowers. In the preface to the first edition of his book, *The Worst Journey in the World*, he sums up the virtues of the three towering figures of Antarctica's Heroic Age: "For a joint scientific and geographical piece of organization, give me Scott … for a dash to the Pole and nothing else, Amundsen: and if I am in the devil of a hole and want to get out of it, give me Shackleton every time."

The Least You Need to Know

- Robert Falcon Scott led a large expedition in 1902–1904 that begin the exploration of Antarctica.
- Ernest Shackleton, who had been on Scott's 1902–1903 South Pole trip returns to Antarctica in 1908–1909 and gets within 94 miles of the Pole.
- In 1909, both Frederick Cook and Robert Peary claim to have reached the North Pole.
- In December 1911, Roald Amundsen and four other Norwegians reach the South Pole.
- Robert Scott and four other Englishmen reach the South Pole a month after Amundsen, but all die on the way back.
- Ernest Shackleton's dream of crossing Antarctica ends in 1915 after ice crushes his ship, the *Endurance*, but all 27 men with him survive.

Chapter 13

Polar Exploration Between the Wars

In This Chapter

- Polar exploration enters a technological age
- Richard Byrd becomes famous for a polar flight
- An airship crosses the Arctic via the North Pole
- Richard Byrd leads the first of three expeditions to Antarctica, beginning an era of mechanized exploration
- Byrd and four others fly over the South Pole

When polar exploration resumed after World War I, private funds still paid most of the expenses, but that started to change before the next two decades were over. The biggest changes after World War I, however, were in technology. No longer would explorers disappear over the horizon for months; now even small parties kept in contact by radio. Probably even more important, faster and more reliable aircraft extended explorers' range. Trips that before World War I took months could now be completed in hours. The 1920s and 1930s saw the United States become the leading nation in the new kind of Antarctic exploration that relied on mechanization and which had a strong scientific focus.

The Heroic Age Gives Way to the Air Age

In *The Worst Journey in the World*, a memoir of Robert Falcon Scott's 1910–1912 Antarctic expedition, Apsley Cherry-Garrard wrote that Scott was "the last of the great geographical explorers … and he is probably the last old-fashioned polar explorer, for, as I believe, the future of such exploration is in the air, but not yet."

Cherry-Garrard was certainly correct about Scott being the last of the great geographical explorers; no large area of land on earth remained undiscovered. Antarctica certainly had many mountains and huge areas of ice no human had yet seen, and people didn't know whether a large island sat in the Arctic Ocean between Alaska and the North Pole. But these explorations were matters of filling in gaps in maps with known outlines.

Cherry-Garrard was also correct about the future of polar exploration being in the air. Within seven years of publication of Cherry-Garrard's book, people would look down on both the North and South poles from the air.

A New Kind of Radio Ends Polar Isolation

After aircraft, the new technology that surely made the most difference, especially in Antarctica, was radio. It's true that Guglielmo Marconi had sent a radio signal across the Atlantic in 1901, and when World War I ended in 1918, powerful transmitters were regularly sending signals across oceans. But 1918 radios worked only at night for long distances. Since almost all polar exploration takes place in summer's continuous daylight, radios were of little use.

A 1925 expedition led by Donald MacMillan ended the days of polar isolation. MacMillan was an experienced explorer who had been on Peary's 1909 North Pole expedition. His second in command was Eugene F. McDonald, a Naval Reserve officer and owner of the then-small Zenith Radio Company. McDonald brought along some of Zenith's new *short-wave* radios. Unlike the radios then in common use, short-wave radio works over long distances day and night; the expedition was in almost daily contact with the expedition's sponsors, the National Geographic Society, and the U.S. Navy in Washington.

Polar Talk

A radio that transmits and receives in the 3 to 30 megahertz frequency range is called a **short-wave radio.** Radio waves in this range are reflected by the layer of charged particles known as the ionosphere, more than 34 miles (55 kilometers) above the earth.

Probably as important as the radios for the future of polar exploration were the three Navy Loening amphibious airplanes with the expedition under the command of Richard E. Byrd, a U.S. Navy officer. Among other lessons, Byrd learned that open-cockpit, single-engine airplanes weren't suitable for polar flying.

Polar Explorers Give Up Dogs for Aircraft

In the early 1920s, Roald Amundsen turned his attention to reaching the North Pole by air. As was often the case, he was short of money and came to the United States to raise funds. Lincoln Ellsworth, a World War I pilot and adventurer who had caught the polar bug, read about one of Amundsen's lectures, and arranged to meet the explorer. Ellsworth's father, James, was a mining millionaire who wasn't eager to see his heir head off to the ice. Lincoln persuaded his father to invest in Amundsen's plans, apparently selling the idea that he would be in good hands with Amundsen in charge.

In early May 1926, Amundsen and Ellsworth came to Spitsbergen with an *airship*, the *Norge* (*Norway*), financed by Ellsworth (who had inherited his father's fortune), and designed and built in Italy by Umberto Nobile, who was to pilot it across the Arctic Ocean.

Polar Talk

An **airship** is a lighter-than-air aircraft with an engine that moves it through the air, and which can be steered. The main body is a huge, cigar-shape balloon filled with a lighter-than-air gas, now usually helium.

Also in Kings Bay that May was Richard Byrd, the U.S. Navy aviator, with a privately financed German Fokker tri-motor (one engine on each wing and one on the nose) airplane, and about 50 sailors and Marines to work on it. Byrd and Floyd Bennett, a Navy pilot, intended to fly to the North Pole and back to Kings Bay. When one of the Fokker's skis broke, Amundsen asked Bernt Balchen, a Norwegian pilot with the *Norge* crew, to give the Americans a hand. Balchen helped fashion skis that would be sturdy enough, showed the Americans how to wax them, and advised Byrd and Bennett to take off at midnight when the sun was low in the sky, which made the snow colder and drier, thus less sticky. They'd have a better chance of getting off the ground in their airplane loaded with fuel for the 1,518-mile (2,442-kilometer) round trip to the Pole.

Byrd and Bennett Head North

A little after midnight on May 9, 1926, Bennett, the pilot, and Byrd, the navigator, took off, heading north. They returned fifteen and a half hours later, saying they had reached the North Pole. That evening, Byrd, Bennett, Ellsworth, and Amundsen had "a splendid dinner," Byrd wrote later. As dinner was ending, Amundsen asked Byrd what he planed to do next. "The South Pole," Byrd answered. "You have the right idea" to use airplanes, Amundsen told Byrd, and offered him some equipment. That evening, the two of them begin planning Byrd's air assault on the South Pole.

Questions Arise About Byrd's North Pole Flight

As soon as Byrd and Bennett landed back at Kings Bay, some began to doubt whether they had actually made it all the way to the North Pole. To complete the trip in only fifteen and a half hours would require a ground speed of around 100 mph (161 kph), considerably faster than the Fokker could go, especially with skis slowing it down. Byrd said they had tail winds going and coming from the Pole. At the time, no one, at least no Americans, expressed doubts publicly; the news went out to the world, and Byrd returned to the United States a hero.

Byrd asked Balchen to come to the United States, where he helped Bennett pilot the North Pole plane on a national tour, went to Antarctica with Byrd, and became one of the nation's leading polar aviators after becoming a naturalized citizen.

Explorations

The question of whether Byrd and Bennett made it to the North Pole in 1926 has been hotly debated since the 1950s, when Bernt Balchen wrote an autobiography in which he said that Bennett, who died in 1928, as much as told him that he and Byrd had been forced to turn back before reaching the North Pole. When word got out what Balchen said in the yet-to-be-published book, Byrd's supporters pressured the publisher not to ship the book to stores and to destroy the copies printed. Richard Byrd, who died in 1957, was a public hero and highly regarded by the Navy. He had powerful friends, including his brother Harry Flood Byrd, who was a U.S. Senator from Virginia from 1933 to 1965.

An Airship Crosses the Arctic Ocean

Amundsen, Nobile, Ellsworth, 13 other men, and Nobile's fox terrier, Titina, took off in the airship *Norge* on May 11, 1926. They reached the North Pole the next day, where they dropped Norwegian, Italian, and U.S. flags attached to aluminum shafts that speared the ice and stayed upright.

From the North Pole, the *Norge* kept on going until bad weather forced it to land on May 14 at Teller, Alaska, near its destination, Nome. They found that the Arctic contains no large land mass between Alaska and the North Pole.

CAUTION

Crevasse Caution

If you're on the TV show *Who Wants to Be a Millionaire* and the question is: "Who was first to reach the North Pole," you could answer: "Frederick Cook and two Eskimos," "Robert Peary, Matthew Henson, and four Eskimos" or "Richard Byrd and Floyd Bennett in an airplane." You might even say, "the 16 men on the *Norge*," You can find someone to vouch for each of these answers. The Scott Polar Research Institute at Cambridge University says it was the *Norge* men, but many people argue that flying over the Pole isn't reaching it.

A Second Airship Expedition Ends in Disaster

In the spring of 1928, Umberto Nobile attempted to land the *Italia*, a sister airship to the *Norge*, at the Pole. The *Italia* took off on May 23, 1928, but found the wind too strong to land at the Pole. The next day it crashed on the ice north of Spitsbergen with nine of the 16 men on board surviving, including Nobile and his dog, Titina. The crash prompted a large, but uncoordinated search involving airplanes and ships from five nations. Roald Amundsen joined the search, but a plane carrying him and five others disappeared, and no sign those aboard was ever found. One of the nine who survived the *Norge* crash died, but the others, including Nobile and his dog, were rescued.

Byrd Opens a New Era of Antarctic Exploration

Until Richard Byrd's first Antarctic expedition of 1928–1930, no American expedition had gone south since 1840, when Navy Lieutenant Charles Wilkes explored about 1,240 miles (2,000 kilometers) of the continent's coast. In the late 1920s, the fame Byrd won with his North Pole flight, his knack for publicity, and the prosperity of the time made raising money for expeditions relatively easy and he led a large, well-equipped expedition to Antarctica in 1928.

The two ships carrying the Byrd expedition arrived at the Bay of Whales on Antarctica's Ross Sea the last week of December 1928 and began unloading on January 2, 1929. Byrd's first big decision was where to build the camp he had decided to name "Little America." Byrd decided on a site about 8 miles from the ships and 4 miles north of where Amundsen's 1910–1912 Framheim camp had been. During the next 3 months, the expedition's Ford snowmobile and dog teams hauled 650 tons (660 tonnes) of supplies and equipment up the slope to the camp. The smaller Fokker Universal and Fairchild airplanes took off from the ice close to the ship and flew to Little America. All 97 of the available dogs pulled the larger Ford tri-motor, named the *Floyd Bennet*, to the camp.

Ice Chips

Even though Little America was only four miles from the site of Amundsen's Framhein base, the Byrd party never managed to find it, even using an instrument to detect metal under the snow. While most of the Arctic is a desert, heavy snow falls along the coast, and from January 1 to September 9, 1929, 93.8 inches fell on Little America. Framheim was most likely buried by snow.

New Technology Doesn't Drive Out the Old

Equipment hauled to Little America included 24 radio transmitters and 31 receivers. These were used on the two ships, the main base, two smaller bases, the three airplanes, and three dog teams at any one time. Five radio engineers were along; a sign of how important radios were to Byrd. "The radio beyond doubt has ended the isolation of this ice cap. As a practical thing, its help is priceless," he wrote in 1930 in *Little America*. "But I can see where it is going to destroy all peace of mind, which is half the attraction of the polar regions."

Byrd Begins True Aerial Exploration of Antarctica

The expedition's pilots began making short test flights on January 15, and on January 27, 1929, Byrd, with Balchen as pilot and Harold June as the radio man, took off in the Fairchild to explore to the east of Little America. In comparing his efforts with those of Heroic Age explorers, Byrd wrote, "we were exploring snow-covered land … at the rate of 4,000 square miles an hour." About 120 miles (190 kilometers) east of Little America they saw a range of at least 14 mountains, which Byrd named the Rockefeller Range. As they neared one on the mountains, Byrd saw bare rock above the ice, some with "brown and black coloration." He knew Lawrence Gould of the University of Michigan, the second in command and expedition geologist, would want to investigate.

Explorers Head South by Dog Sled and Airplane

In preparation for an attempt to fly to the South Pole, the expedition's Ford tri-motor flew 440 miles (708 kilometers) south along Amundsen's 1911 route to the bottom of the Liv Glacier where those aboard, including Byrd, set up a depot with gasoline, oil, and about 300 pounds of food on November 18. The tri-motor would be unable to fly to the Pole and back without stopping for gas; this depot would be the required stop.

On the way south, Byrd asked himself the question Antarctic pilots ask today as they head out to land in a place, where no one has groomed a *skiway:* "What would we find at the foot of the mountain? A firm, smooth surface? Or a surface torn and marred by crevases and *sastrugi?*" Fortunately, Byrd's airplane found a good landing place.

Amundsen's Three-Month Trip Takes Less Than a Day

After a radio report of clear weather from a geological party 400 miles to the south on November 29, 1929, Byrd, who navigated; Balchen, the pilot; June, the radio man; and Ashley McKinley, the photographer, took off in the Ford tri-motor for the Pole. They were following Amundsen's route. "Amundsen was delighted to make 25 miles per day," Byrd wrote. "We had to average 90 miles per hour to accomplish our mission." The speed and comfort of the 1929 journey was offset by "an enlarged fallibility. A flaw in a piece of steel, a bit of dirt in the fuel lines or carburetor jets … could destroy our carefully laid plans …"

Polar Talk

A **skiway** is a runway of groomed snow used by ski-equipped airplanes at bases and temporary research camps where planes will be coming and going.

A **sastrugi** is a long, wave-shape ridge of hard snow formed by the wind. They are common in the polar regions.

The big hurdle was climbing high enough to get over the mountains that bound the Polar Plateau. Amundsen's party had climbed the Axel Heiberg Glacier, which topped out at 10,500 feet (3.2 kilometers) above sea level, which is about as high as the Ford was designed to fly. They followed the nearby Liv Glacier, which seemed wider and not quite as high. On the way up they saw ice falls—frozen waterfalls 200 to 400 feet (60 to 120 meters) high. "Beautiful yes," Byrd wrote, "but how rudely and with what finality they would deal with steel and *duralumin* that was fated to collide with them at 100 miles per hour."

Polar Talk

A strong, lightweight alloy of aluminum and copper, **duralumin** is used for aircraft parts. The term is now obsolete in the United States.

To make it over the mountains, they had to get rid of weight: two bags of food, 250 pounds, splattered on the ice below, enough to feed the four of them for a month if they had been forced down. From here on they couldn't really relax, but the worst was over. The four men on the plane became the first to see the Pole since Scott's doomed party had left in January 1912; they turned back toward the coast, refueled at the depot, and landed back at Little America 18 hours and 41 minutes after leaving.

Richard Byrd before leaving for the South Pole flight. He dropped the American flag at the Pole. It's weighted with a rock from the grave of Floyd Bennett, his North Pole pilot.

(Byrd Polar Research Center Archival Program)

Byrd Returns to Answer Scientific Questions

When Byrd and his men arrived back in the United States in June 1930, they were heroes, and Byrd was promoted to admiral. Byrd's book, *Little America*, became a best-seller. He closed it with: "The Antarctic has not been conquered. At best we simply tore away a bit more of the veil which conceals its secrets … It still remains, and will probably remain for many years to come, one of the great undone tasks of the world." Byrd was eager to return, but the world was deep into the Great Depression and raising money for a new expedition delayed departure until 1933. Even so, the expedition that headed south was even larger and better equipped than the last one.

Byrd Almost Dies Alone on the Ice

One of the expedition's goals was to study Antarctica's weather. In order to collect inland data, an "advanced weather base," was set up 123 miles (198 kilometers) south of Little America with the plan that three men would spend the winter there. But delays made it impossible to haul enough food for three men to the 9 by 13-foot (2.7 by 3.9-meter) hut they would live in. Byrd figured that two men wouldn't be able to stand each other for the entire winter there. It would have to be one, and because Byrd didn't want to ask someone else to do it, he settled in alone on March 28, 1934.

Richard Byrd in his hut shortly after rescuers reached him in August 1934.

(Byrd Polar Research Center Archival Program)

By late May when the temperature had dropped to –96°F (71°C), frozen condensation partly blocked the hut's ventilators, allowing carbon monoxide from the stove and generator engine to build up in the hut. Byrd became deathly ill and thought he was going to die, but managed to radio in weather observations on schedule three times a week. He didn't say anything about his illness because he didn't want men risking their lives to rescue him. But Byrd's radio messages showed something was wrong and three attempts were made to reach him with tractors before three men reached him in a tractor on August 10, 1934.

Ice Chips

Reaching Byrd with a Citroen tractor was just one more sign that motorized vehicles could work in Antarctica. Tractors were used on one 230-mile (370-kilometers) depot-laying trip when it was too cold to use dogs. The days of dogs were numbered, although their barks would echo over Antarctica's ice until 1993, when they were banned because of a fear they could spread distemper, a canine disease, to seals.

Science and Publicity Mark the Expedition

The 1933–1935 expedition answered the last big geographic question about Antarctica by establishing that there was no strait connecting the Weddell and Ross seas; Antarctica is a single continent. Scientific accomplishments included measuring the depth of the ice cap, discovering that more forms of life than anyone expected live in the ocean around Antarctica, and collecting huge amounts of data on cosmic rays, weather, and geology.

In addition to around a dozen scientists, the expedition included Charles Murphy, a CBS Radio correspondent, who helped produce weekly radio shows from Antarctica for the CBS network. Communications from Little America during the 1928–1930 expedition were via the dots and dashes of Morse code because no short wave radio of the time could broadcast high-quality voice signals. But by 1933, the new Collins radio could send commercial-broadcast-quality talk and even music from Antarctica. Beginning February 3, 1934, depression-era Americans gathered around the living room radio each Saturday night to listen to broadcasts from Antarctica, sponsored by General Mills, the makers of Grape Nuts, the expedition's official cereal.

The U.S. Government Goes Exploring in Antarctica

The 1933–1935 Byrd expedition had stirred even more interest in Antarctica and the government began to take notice. On January 7, 1939, President Franklin D. Roosevelt issued a memorandum approving plans for a U.S. Antarctic Service which would send an expedition south under Byrd's command.

The president's decision involved more than a love of exploration. Other nations, including Hitler's Germany, were asserting claims to parts of Antarctica based on exploration. The U.S. position since 1921 had been that such claims weren't enough; claiming sovereignty required actually "settling" the land. The two U.S. Antarctic Service bases would have Americans there all year from 1939 on. Expedition flights mapped about 700 miles of Antarctic coast and continued to fill in details of the continent's mountains, bays, glaciers, and ice sheets. Scientists collected auroral, biological, cosmic ray, magnetic, and seismic data.

With World War II already raging in Europe and Asia, and the United States likely to become involved, Congress refused to appropriate money for another year, and the last Americans left Antarctica in March 1941. Antarctica was put aside until the war was won.

The Least You Need to Know

- Beginning in the 1920s, short-wave radios ended the isolation of polar explorers.
- Also beginning in the 1920s, aircraft extended the range of explorers.
- In 1928, Richard Byrd led a large expedition to Antarctica, the first American expedition since 1840.
- In 1929, Byrd and three others flew to the South Pole without landing.
- In 1934, during Byrd's second expedition, carbon monoxide almost killed him as he manned a remote weather station alone.
- In 1939, the U.S. Antarctic Service was formed and Byrd led its large expedition to Antarctica.

Chapter 14

Big Science Comes to the Polar Regions

In This Chapter

- The Cold War makes polar regions militarily important
- Arctic science takes to the air and ice
- The United States returns to Antarctica in a big way
- Polar science becomes internationalized
- The legacy of the International Geophysical Year

Polar exploration resumed after World War II, but with military and political concerns playing a larger role than before. During the Cold War, the United States and the Soviet Union faced each other across the Arctic. Even during the Cold War, however, scientists conceived the global program of intensive, global scientific observations known as the International Geophysical Year (IGY). In 1957 and 1958, the IGY brought unprecedented numbers of researchers to both the Arctic and Antarctic, and changed the nature of polar exploration and science. Since the late 1950s, thanks in large part to the IGY, stations, equipment, and air transportation have made today's wide-ranging polar science possible.

The USA and USSR Turn Their Eyes to the Arctic

While the nations that fought World War II had some strategic interests in the polar regions, the only major combat was German attacks on allied ships carrying war supplies to Russian ice-free, Arctic Ocean ports such as Murmansk. Soon after the war ended, it was becoming clear that the United States and the Soviet Union—separated by the Arctic Ocean—were opponents on the world stage. "If a third world war breaks out," General Henry "Hap" Arnold, the retired Army Air Force chief, said in 1946, "its strategic center would be the North Pole." The U.S. military had to learn to live, fly, and fight in polar regions.

In pursuing these goals, the Navy conduced cold weather exercises in the summer of 1946 off Greenland, and established the Naval Arctic Research Laboratory in Barrow, Alaska. The Air Force launched "Project Nanook," which had the stated goals of searching for undiscovered Arctic land, learning more about Antarctic weather, and improving polar navigation. The reconnaissance squadrons involved were also—no surprise—keeping an eye on the growing Soviet air force on the other side of the Arctic.

Ice Chips

Paul Siple, who started his polar career as a Boy Scout with Byrd's 1928–1931 expedition, became one of the nation's leading polar scientists. On October 8, 1946, he flew on the first Air Force B-29 to fly over the North Pole. "The flight took twenty-four hours for the simple reason that we kept straying off course," he wrote. "It was obvious … that there was much to be learned" about polar navigation.

Scientists Go with the Floe

In late July 1950, the radar on an Air Force reconnaissance airplane spotted a "target" about 500 miles (800 kilometers) northwest of Barrow, Alaska, that looked like land on radar, and was named "T-3" for the third such "target" found. T-3 was an "ice island," a chunk of ice about 9 miles (14 kilometers) long and 4.5 miles (7 kilometers) wide. It had broken off from an Ellesmere Island ice shelf, much like the huge icebergs that break off from Antarctica's ice shelves.

In March 1952, an Air Force C-47 transport landed on T-3, which was then about 103 miles (166 kilometers) from the North Pole. Colonel Joseph O. Fletcher, commanding officer of the 58th Weather Squadron, and two other men set up camp. This was the beginning of three decades of on-and-off U.S. use of drifting ice stations, including on T-3, which came to be known as "Fletcher's Ice Island." The Americans didn't know it until later, but Soviet scientists had started studying the Arctic from drifting stations in 1937. This work continued until 1991 when the Soviet Union collapsed.

The United States Returns to Antarctica

Thanks in part to the urging of Admiral Richard Byrd, the U.S. Navy returned to Antarctica in 1946 with the largest expedition ever sent there. Operation Highjump consisted of 13 ships, including a submarine and an aircraft carrier, numerous aircraft, and 4,700 men. One of the Navy's goals was to learn how to operate in polar waters—fleet maneuvers this large in the Arctic would surely alarm the Soviets. Another goal was to continue the mapping of Antarctica begun before World War II. A final aim, according to the orders for Admiral Richard Cruzen, the commander, was to consolidate and extend "United States potential sovereignty over the largest practicable area of the Antarctic Continent."

Operation Highjump, during the Southern Hemisphere summer of 1946–1947 with no one spending the winter, managed to map about 1.5 million square miles of Antarctica using Navy flying boats, which took off and landed on the water, and even R4D Skytrains (the Navy version of the DC-3), which flew to a new Little America base from the aircraft carrier USS *Philippine Sea*, from about 700 miles off shore to become the first aircraft to land on Antarctica after taking off somewhere else.

Science Becomes the Reason to Head South

When the Cold War was at one of its hottest points in the beginning of the 1950s as hot war raged in Korea, scientists around the world began putting together a dream. In 1952, the International Council of Scientific Unions proposed 18 months of intense observations of the earth from July 1957 through December 1958.

The scientists picked this period because the sun would be at a peak of its 11-year cycle, when, among other things, auroras occur more often and farther from the poles. All the earth would be observed during the International Geophysical Year (IGY), but the polar regions would be a focus.

Sure, We'll Build a Station at the South Pole

During planning for the IGY, some scientists began talking about setting up a station at the South Pole. In fact, some wanted not only a station, but one with men there making scientific observations through the winter. "The path to the South Pole is fraught with great dangers," Paul Siple told the National Academy of Sciences, sponsors of the U.S. part of the IGY. Siple doubted that tractor trains could haul enough material to build shelter, and the needed fuel, food, and other supplies for even a small group to spend the winter.

Paul Siple during one of the 1930s Antarctic expeditions.

(Byrd Polar Research Center Archival Program)

When Siple went to the IGY's first international planning conference in Paris in July 1955, the United States had not decided about a Pole station. Then, "the Russians dropped a bombshell," Siple wrote. They planned to build and man a South Pole station. The Americans were speechless. Then the French conference chairman said, "I'm sorry, but we have accepted the offers of the United States to erect and man a South Pole Station. We don't think there should be two there."

The Americans couldn't back down from such a commitment, even though a Frenchman had made it. "Somehow we would have to erect a self-sufficient village at the Pole to house and support a group of Americans," Siple wrote. "It promised to be the most difficult construction job in history." The military, the U.S. Navy, was the only organization that could do this job.

Navy Pioneers a Base and a Way South

But by the mid 1950s, movements of the Ross Ice Shelf's eastern end were making it clear that Little America was not a good place for a station. McMurdo, which had been envisioned as a South Pole supply base, became the main U.S. station, as it is today.

The Navy Worries About Flying to the Pole

While other U.S. stations were built and supplied for the coming winter, the centerpiece of the 1956 and 1957 summer was building, supplying, and manning the South Pole station. Only ten men had ever set foot there—Roald Amundsen and his four men in December 1911, and Robert Scott and his four in January 1912. Byrd and his plane crews had flown over the Pole in 1929 and again during Operation Highjump, but no one really knew whether an airplane could land there and, even more important, take off again.

The Navy set up a refueling base on the Ross Ice Shelf about 390 miles from the pole. The Air Force had eight C-124 Globemaster transports at McMurdo to parachute supplies for the South Pole Station. When the first Navy plane landed at the Pole, one of the Globemasters was overhead, ready to drop emergency supplies, sleds, and a small *weasel* tractor, if the Navy plane couldn't take off. The Navy was even prepared to parachute two teams of 11 dogs to a stranded crew. Whether with weasels or dogs to pull the sleds, the stranded men would head for the refueling base.

Polar Talk

A **weasel** was a World War II tracked vehicle designed to haul soldiers and equipment over a variety of terrain, including snow. It was about 10 feet long and 5 feet wide, and weighed about 4,000 pounds (1,814 kilograms). Instead of wheels, it had treads that ran the vehicle's length on both sides.

At 8:45 P.M. on October 31, 1956, a Navy R4D airplane named *Que Sera Sera* ("what will be, will be") landed on skis at the South Pole and Admiral George Dufek stepped out, becoming the 11th person to leave his footprints in the Pole's snow. "The bitter cold struck me in the face and chest as if I had walked into a swinging door," Dufek wrote later. The temperature was –50°F (–45°C).

The pilot, Lieutenant Commander Gus Shinn, described what happened after Dufek gave the order to leave after 49 minutes on the ice: "I pushed the throttle fully forward and nothing happened. I then fired four *JATO bottles* and still nothing. I fire another four and got slight movement, then four more, followed by another three and we more or less staggered into the air."

Polar Talk

JATO bottles, or "Jet-Assisted Take Off" rockets, provide added thrust for airplanes to take off on short runways, at high elevations, or in snow. While the name JATO has been used since World War II, they are really rockets, not jet engines, which is why the 109th Airlift Wing, which uses them on its LC-130 airplanes in Antarctica and Greenland, uses the term "ATO" for Assisted Take Off.

A Special Breed Builds the Pole Station

Paul Siple, who was to be in charge of the eight civilian scientists who would spend the winter at the Pole, and Lieutenant (jg) Jack Tuck Jr., commander of the eight Navy men who would spend the winter, wanted construction to begin right away. But it didn't start until two R4Ds landed at the Pole in bright sunlight at midnight on November 20, 1956, with Lt. Richard Bowers, his team of seven Navy *Seabees*, and 11 sled dogs to begin construction.

Polar Talk

Sailors who are skilled in construction trades and members of construction battalions are CBs or **Seabees.** They first gained fame during World War II for building airfields and other facilities on captured Pacific islands while the bullets were still flying.

The first few days the men lived in survival tents, but they quickly erected two *Jamesway* huts to live in as they built the station. Air Force Globemasters dropped by parachute most of the materials needed to build the base and supplies that would keep the men alive all winter. Unfortunately, about one-fifth of the chutes failed to open.

Polar Talk

A **Jamesway** is a prefabricated, canvas building shaped like a half circle with a wooden floor and frame, which is easily erected and taken down. Many dating back to the time of Operation Deepfreeze I are still used at U.S. Antarctic bases and field camps.

Siple arrived on November 30, and his description of the next two months is one tale after another of materials or supplies ruined when they "streamed in" under unopened parachutes. At times the wind would drag functioning parachutes and the things attached to them 20 miles from the station. The parachutes were a good source of cloth used for many things, such as decorating rooms (they were different colors), and as bags filled with snow and dragged inside to melt for drinking water.

Explorations

On January 1, 1957, the 16 men who were building the South Pole Station, buried a time capsule to be opened in 2000. Paul Siple, the station's chief, ruled against burying it in the snow because "it never would be found again." It went into the small plywood shack that marked the South Pole. In 2002, "We don't know where it is," Jerry Marty, the National Science Foundation representative at the Pole, told the Antarctic Sun. A new capsule, which was put into one of the support beams of the new station, is being better documented so it can be found for its 2050 opening.

On February 21, 1957, after the last of the Seabees had left, the 18 men at the Pole knew they wouldn't see another human until late October because it would be too cold for an airplane to land. Siple described the group "as 18 men in a box" they had to stay in to survive. They had no "possibility we could be rescued should tragedy strike." The nearest humans to the 18 men at the South Pole were three men wintering over at the British South Ice Station, about 560 miles (900 kilometers) away.

The IGY Begins with a Blood Red Sky

Scientists had picked the IGY dates to coincide with a period of high solar activity. The sun didn't disappoint them. July 1, 1957, brought one of the most severe solar storms of the year, Siple writes: "... we had our most spectacular red aurora ... So with the horizon blood red in the enveloping darkness and the electrical storms raging above us, we were well begun on the IGY program."

The effects of the solar storm, of course, were being observed not only at the nearly 60 stations in Antarctica, but at others around the world, including more than 76 locations in the Arctic—some on ice islands where U.S. scientists were making observations.

Pole Remains Isolated After the Sun Returns

The sky began to show the first sign of the sun in August 1957, but civil twilight, when the men could move around outdoors without flashlights, arrived so slowly the men at the South Pole could hardy say when it started. "By September 6," Siple wrote, "the dull grays of the sky were interlarded with other colors." The sun came up on September 23, 1957, but it was still bitterly cold, the South Pole had set a new low temperature record for the world of –102.1°F (–74.5°C) on September 18, 1957.

The first plane landed on October 21, but broke down, probably from the cold, and its 16 men joined the 18 at the Pole until November 16, when airplanes were able to arrive and depart, taking the 18 men who had spent the winter to McMurdo, and bringing replacements. Since that first winter, the South Pole has never been unoccupied.

Long-Distance Exploration Resumes

During the Antarctic winter scientists were more or less stuck in their "boxes" around the continent, not just at the South Pole. With the return of the sun and warmer temperatures, scientists and those who are needed to keep scientists in business in the polar regions, such as tractor mechanics, set out on several traverses across the ice.

These trips, using tractors of various kinds, often with air support, were much more than attempts to get somewhere, such as Amundsen's and Scott's trips to the South Pole in 1911 and 1912. While some of these earlier trips collected a little scientific data, that was a sidelight. The IGY traverses did such things as measuring how long the sound from a small explosion set off in the ice took to reach the rock below the ice and bounce back—thus measuring the ice's thickness.

Land traverses during the IGY.

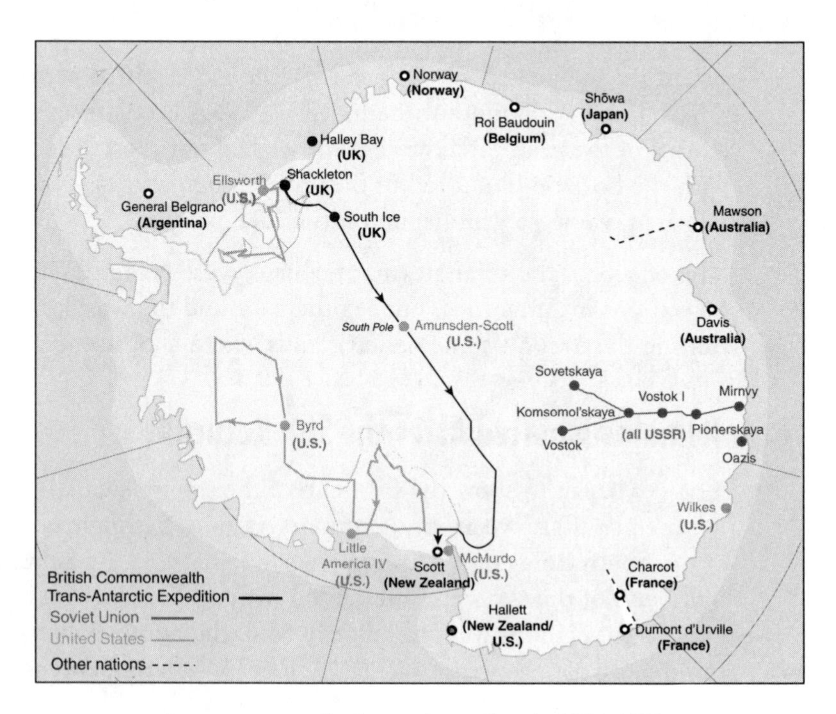

Journeys across the ice, major stations in 1957–1958

Shackleton's Dream Is Finally Realized

The British Commonwealth Trans-Antarctic Expedition of 1957 accomplished what Shackleton had set out to do in 1915 when the ice crushed his ship, the *Endurance*.

Edmund Hillary, who had become famous in 1953 when he and Tenzing Norgay were the first to climb Mount Everest, led New Zealand's part of the joint expedition with Great Britain, which aimed to cross the continent.

Early in 1957, he established New Zealand's Scott Base on Ross Island, just down the road from the U.S. McMurdo Station. Hillary's groups headed toward the South Pole in October 1957, setting up supply depots for a British party that would be coming from the other side of the continent. His party, using Ferguson farm tractors, reached the South Pole on January 3, 1958.

Meanwhile, on November 24, 1957, a party led by the British explorer Vivian Fuchs had left Shackleton Base on Vahsel Bay on the Weddell Sea. This group was using four sno-cats (large tracked vehicles), three weasels, a Muskeg tractor, and men with a dog team to scout the best route. They reached the South Pole on January 18, and then went on to Scott Base, arriving on March 2, 1958, becoming the first people to cross the continent, a trip of 2,158 miles (3,472 kilometers).

Ice Chips

[N]o doubt about it—our tractor train was a bit of laugh. But despite appearances, our Fergusons had brought us over 1,250 miles (2,000 kilometers) snow and ice, crevasse and sastrugi, soft snow and blizzard to be the first vehicles to drive to the South Pole.

—Edmund Hillary, upon arrival at the South Pole

The Legacy of the IGY Lives at the Poles

From 1955 through 1957, as the United States built the South Pole Station and began turning the McMurdo Station into today's largest Antarctic "city," those involved charged ahead with a lot of confidence—and also with a lot of bickering and bureaucratic snarls that are always part of large undertakings. Navy Captain Douglas Cordiner expressed the confidence of the times when he said: "With (Air Force) support and the help of the Navy, I think we could put igloos in hell if the IGY program called for it."

Those who conduct polar research today continue to benefit from the legacy of the IGY. In July 2000, Rita Colwell, director of the National Science Foundation, said the IGY "led scientists to come to think of the poles as natural laboratories in which to capture and integrate diverse data about the heavens and the earth." In Chapters 20 through 28, we will see how polar science has grown since the IGY, and how the questions it is asking have become more and more important to everyone on Earth.

The Least You Need to Know

- The largest U.S. Antarctic expedition was the Navy's Operation Highjump in 1946–1947 with 4,700 men.
- The International Geophysical Year intensely observed the earth in 1957–1958, with a strong polar focus.
- Those aboard a U.S. Navy airplane that landed at the South Pole in 1956 were the first people there since 1912.
- The U.S. South Pole Station, built in 1956, has been continuously occupied since then.
- From October 1957 through March 1958, a British–New Zealand expedition crossed Antarctica on tractors, becoming the first people to cross the continent on the surface.
- The 1957–1958 International Geophysical Year laid the foundations for today's polar science.

Part 4

The Polar Regions Today

Even though the polar regions are far from the rest of the world, they do not escape all its problems, and they are not as hard to reach as in the past. In this part, we'll first look at how the indigenous people of the Arctic, especially in North America and Greenland, are coping with conflicts between their traditional life styles and the larger world.

Then, we'll head south to look at Antarctica today, including the tangle of international rivalries and diplomacy that led to the Antarctic treaty. We will then turn to day-to-day life in Antarctica and look at how people travel, live, and work—including telling you how to find an Antarctic job. We'll close this part by giving you information on planning a polar trip that fits your budget and spirit of adventure.

Chapter 15

Today's Life in the Arctic

In This Chapter

- Today's Arctic life combines old and new ways
- Alaska's Eskimos run companies, hunt whales
- Canada sets up an Inuit territory
- Inuits from four nations come together
- Oil drilling is the big issue in Alaska

In this chapter, we look at the conflicts between traditional and modern life in the parts of the Arctic that are the home of the Inuit: Alaska, Canada, Greenland, and far-eastern Russia, where a tiny outpost of Inuit culture managed to survive communism. We'll see that traditional and modern ways are thoroughly mixed across most of the region. Some native people have done quite well in the modern world while others struggle.

Life in the Arctic Is Different

When you visit Barrow, Alaska, you know you're still in the United States. The 50 or so channels on the television in your hotel room are mostly those you watch back home, and you pay in dollars for a traditional American breakfast of bacon and eggs. But during an early morning stroll down the gravel street past the airport to the Arctic Ocean beach, you see two bearded seals, which someone has started to butcher, on a piece of plywood next to the street.

Ice Chips

When sea ice broke away from land near Barrow, Alaska, in March 2002, threatening to take 18 seal hunters out to sea, some hunters radioed for help. Some used Global Positioning System devices to help North Slope Borough Search and Rescue service, and the Barrow volunteer Search and Rescue service helicopters find them.

The mix of traditional ways of living and the twenty-first century Western world is probably more apparent in Barrow than in many other towns and villages around the Arctic. Still, hotels in tiny Inuit villages in the Canadian Arctic advertise cable television and computer hookups, and visitors arrive by air, not the scurvy-ridden ships that brought the first Europeans to these areas.

Many Inuit live traditional lives today in the sense that they rely on hunting sea mammals or caribou for an important part of their diet. When hunting, however, they're likely to travel by snowmobile, not dogsled, and they use rifles, not bows and arrows.

You will not find an Inuit version of Colonial Williamsburg in the Arctic with men and women in costumes pretending to live in the old ways. On the other hand, you will discover that many of the old ways are more a part of everyday life for large numbers of people than anywhere else in the United States, Canada, or Europe. For instance, the skins of the bearded seals next to the street in Barrow will be used to make a traditional whale hunting boat, and the seal meat and oil will help feed a family.

Alaska's Natives Reach a Land Agreement

In the 1960s, native groups around Alaska organized to press their land ownership claims based on "aboriginal use and occupancy." Well-educated natives used the political and legal systems, and people from different tribes joined together. Native claims cast a shadow on the ownership of almost all land in Alaska, including land that might hold oil and other mineral riches.

A long series of battles ended in December 1971 when President Richard Nixon signed a bill that wiped out native claims to most of Alaska. In exchange, the natives received about one-ninth of the state's land plus $962.5 million in compensation. The money went to native corporations, not to individuals or "tribes."

Some of the native corporations have faltered, while others have done quite well. In 1998, the *Juneau Empire* newspaper looked at what had happened since 1971, and found that the Arctic Slope Native Corporation, headquartered in Barrow, was the most successful. It is a global supplier of oilfield services, which owns other companies around the United States involved in activities ranging from construction to communications.

Barrow also has its village corporation, the Ukpeagvik Inupiat Corporation, which has also been successful. (*Ukpeagvik* is the Inuit name for "Barrow.") The corporation owns a hotel, a construction company, an engineering firm, and the former Navy Arctic Research Laboratory in Barrow, which is now the home of a community college and several scientific ventures.

Ice Chips

The Inuit in the Barrow, Alaska, area are Inupiat, and unlike Inuit in Canada and Greenland, they call themselves *Eskimos*. In fact, two of the companies owned by the Arctic Slope Native Corp. have the word *Eskimo* in their names.

Alaska's North Slope Borough, which is the same as a county in other U.S. states, was formed in 1972 to bring government services to the few people who lived there, using taxes on the oil facilities at Prudhoe Bay. The Borough, which covers about 89,000 square miles (230,000 square kilometers), stretches across northern Alaska from Point Hope on the west to the Canadian border on the east.

The millions in taxes the Borough has collected from Prudhoe Bay brought Barrow and the few villages scattered across the Borough up-to-date schools, health-care facilities, and municipal services never before seen in northern Alaska. But as the twenty-first century begins, the Borough's revenue is falling because the value of the property that's being taxed is declining.

Barrow: Native Traditions Include Science

Barrow, Alaska, has a tradition of science that goes back to the U.S. Navy's Arctic Research Laboratory, which operated there from 1947 until 1980. During that time, many Barrow residents worked for the lab, including managing support activities on some of the "ice island" stations in the Arctic Ocean.

After the Navy closed the lab, the town's Ukpeagvik Inupiat Corporation persuaded the U.S. government to turn the buildings and other facilities over to it instead of demolishing them. Today the old Navy Lab is the center of science in northern Alaska. The Borough's wildlife department conducts research there, and it houses the Borough's community college. It's also the location of the Barrow Arctic Science Consortium, which has an agreement with the National Science Foundation. The Consortium not only brings researchers to the area, but arranges for local students to work with them.

Glenn Sheehan, the Consortium's executive director, says one of the reasons the people of Barrow set up the consortium was to "provide opportunities to local young people and students to work with scientists, something that all of the older community members had experienced and valued highly."

Whaling Is Key to Inupiat Culture

As we saw in Chapter 10, the Inuit development of whale hunting brought more food security to the people living around the Arctic Ocean around 1,000 years ago, until climatic cooling ended whale hunting in most of the Arctic, except on Alaska's northern coast.

"The culture (on the North Slope) is focused around whaling, the harvest, sharing that harvest," says Glenn Sheehan. New England whalers nearly killed all the whales and brought disease in the nineteenth century. "The fact that people are still whaling today is testimony to how important whaling is," Sheehan says. "In spite of all of the death and destruction, they never lost their focus."

Ice Chips

Unlike in the past, people don't starve today if the whale hunt fails. In 2002, Point Hope had a poor season, so the crews there came to Barrow where the local captains let them use boats and equipment, and the Barrow and Point Hope crews split the harvest. The Point Hope whalers took their share of the meat back as checked luggage when they flew back home.

Canada and Its Inuits Settle on a New Territory

By the 1960s, oil exploration in Canada's Northwest Territories made Inuits realize that they didn't control their traditional lands. As in Alaska, a series of political and legal struggles led to a settlement. In 1992, almost 85 percent of the Inuits who would benefit approved a proposed land claims agreement, and by June 1993, the Canadian government turned over control of 135,000 square miles (350,000 square kilometers) of land to the Inuit, which led to the establishment of the new territory of Nunavut in 1999.

Ice Chips

Eighty-five percent of Nunavut's population is Inuit. While Canada's two official languages are English and French, Inuktitut and English are Nunavut's primary languages. In Inuktitut, *Nunavut* means "our land."

Nunavut is the third of Canada's three sub-arctic and arctic territories (the Yukon and Northwest Territories are the other two). Cleaved from the eastern part of the Northwest Territories, Nunavut is continental in size, 800,000 square miles (2.1 million square kilometers). But only around 28,000 people live there. Nunavut's "Big Apple" and capital, Iqaluit, which used to be called Frobisher Bay, has only 4,500 people.

Great Britain transferred ownership of the Arctic Archipelago to Canada in 1880, and most of the 28 far-flung communities scattered across Nunavut today were formed after 1880. Most of the hamlets grew around trading posts, police and military posts, or where missionaries were working. Some places were settled because Canada wanted to assert its sovereignty over remote lands. One theory of international law is that to legitimately claim sovereignty over a territory, a nation has to have people living there. By the 1960s, nomadic Inuit camps had all but disappeared in favor of permanent settlements across Canada.

Besides tourists, several hundred scientists and research assistants swell the population of Nunavut during the summer months. Fanning out over the northern landscape, scientists live in isolated field camps reminiscent of early explorers. And as always, scientists who set up camps along the coast have to remain vigilant; otherwise, they might end up as part of the Arctic food chain by supplementing the protein intake of man-eating polar bears.

Ice Chips

Alert, Canada, is the northernmost place on Earth where people stay, but because it's a military base, weather station, and research station, you wouldn't want to call it a town or village. This leaves Ny Alesund in Norway's Svalbard Islands as the northernmost village. Ny Alesund was called Kings Bay in the 1920s when several attempts to reach the North Pole by air left from there, as we saw in Chapter 13.

Greenland Is Not Quite Independent

Because the ice sheet covers most of Greenland, we might think that calling the place "Greenland" was some kind of real estate scam. In a way, it was: Erik the Red, a Viking from Iceland, used the name to attract the 500 or so settlers he led there in the year 986. The North Atlantic climate was warmer then, during the *Medieval Warm Period*, and settlers in Greenland did well enough to trade with Iceland and Europe through the thirteenth century. As far as anyone knows, the Norse in Greenland were unable to cope with the turn toward a colder climate in the fourteenth century.

Polar Talk

The time believed to have lasted from the eighth through the twelfth centuries when temperatures, especially in the Northern Hemisphere, apparently averaged a few degrees higher than during the preceding and following periods is called the **Medieval Warm Period.**

Links to the long-gone Vikings led Denmark to make Greenland a colony in the eighteenth century. Under today's "Home Rule Government," Greenland's Parliament runs the country while Denmark handles foreign affairs.

Fishing Is Greenland's Big Export Industry

Fishing, especially for prawns (shrimp), is by far Greenland's largest source of foreign exchange. Greenland needs money to import most of the manufactured goods its people need. Most of Greenland's trade is with Denmark, with Greenland importing around $500,000 worth of goods each year but exporting only around $290,000 worth. Denmark covers this deficit with direct grants.

Ice Chips

Greenland sees tourism as a potential source of the foreign exchange the country badly needs. Air Greenland has regular jet service between Kangerlussuaq and Copenhagen, Denmark, but regular flights to and from North America ended in 2001. In January 2003, Greenland's premier Hans Enoksen, said, "We realize that it is crucial to reestablish the airline service to the west if we want to expand our relations and increase our trade with the North American continent."

Around 60 percent of Greenland's 56,000 people live in the 6 largest towns, including Nuuk, the largest town and also the capital. Although the towns have roads, Greenland is like Arctic Alaska and Nunavut, Canada, in that no roads connect the towns. Because all the towns are on or very near the coast, ships and boats haul most goods as well as some people. Also, if you want to go out of town, you generally take an airplane or, for short trips within Greenland, a helicopter.

Explorations

Press commentary in Europe and North American about *A Farewell to Greenland's Wildlife,* a book by Kjeld Hansen, plus reports of hunters killing 40 killer whales in Disko Bay led to hundreds of protest e-mails to Greenland's government in 2002, including threats to drop Greenland travel plans. Terms used in the reports and commentary included *Grimland, wildlife holocaust,* and *killer Inuit.* In response, Greenland's cabinet vowed to improve wildlife management. Greenland "has achieved recognition of the right of people in Greenland to harvest the country's wildlife," the cabinet statement said. "However, we are now facing fierce opposition in the international press because of problems related to our wildlife management. This criticism may have disastrous implications for our society. We therefore have to take very concrete steps to ensure truly sustainable use of our wildlife."

The nature of Greenland's economy means that large numbers of people have to hunt seals, walruses, a few whales, birds, reindeer, and musk oxen, either as their major source of food or to supplement food brought with income from a job.

Russia's Inuit Returning to Traditional Ways

American and other Inuit are helping their cousins across the Bering Strait from Alaska in Russia's Chukotka Region return to some of their traditional ways, years after the Soviet government forced their grandparents to join the march toward Socialism. When the Soviet Union fell, the false economy that had been keeping the Chukotka alive with subsidies vaporized.

Before the 1930s, the Chukotka Inuit had been whale hunters. The Alaska Eskimo Whaling Commission stepped in to help the Russian Inuit by giving them some of their whale quota, sending equipment, and having some Russians come to Alaska to re-learn the old traditions by working with whaling crews.

Other Americans are also helping. The Barrow Science Consortium makes it possible for American scientists to work in Chukotka and has helped Russian organizations do the kind of work it does in Barrow.

Oil Helps Bring Inuit Together Across Borders

"Oil and gas are no new things among us," Eben Hobson, the founder of Alaska's North Slope Borough, told a Canadian hearing on Arctic oil development in 1976. "There are oil seeps throughout our region, and on our way to our hunting camps, we would cut oil-saturated tundra into blocks. Returning from camp in the fall, we would collect these bricks of congealed pitch and tundra and burn them much the same way that urban homeowners use artificial particle logs sold in supermarkets for their fireplaces."

In the 1970s, oil became a major issue around the Arctic when development of the huge oil field at Prudhoe Bay, on Alaska's Arctic Coast, began. Oil companies had proposed similar oil development in Canada's northwestern Arctic, and there was talk of drilling in the Arctic Ocean, and in Greenland or under the ocean off Greenland. "Our big concern, of course, is off-shore development and its threat to our food chain," Hobson told the hearing.

Environmental and other common concerns, prompted Hobson to call for a meeting of representatives from Inuit around the Arctic to meet in Barrow in June 1977. Concerns about oil drilling and other issues led those attending to form the Inuit Circumpolar Conference. Today it looks out for the interests of the estimated 40,000 Inuit in Alaska, 30,000 in Canada, 40,000 in Greenland, and maybe 1,000 in Russia.

Explorations

Inuktitut, the language of the Inuit, is spoken from Alaska to Greenland, but with different dialects and accents, sometimes making it hard for people from different places to understand each other. For instance, attendees at a 1999 Inuit Circumpolar Conference's Language Commission meeting used English to make sure everyone understood each other. Until the nineteenth century, Inuktitut was spoken, but not written. Today it's written in both the Roman alphabet and a script based on the Pittman shorthand method, invented in 1894. The language is most widely used in Greenland where the Greenlandic form is, along with Danish, one of the languages taught in all schools. As in most European nations, many people in Greenland also speak English.

Wildlife Refuge Is the Big Battleground

Over the last few years, if you saw the word *Arctic* in a headline in a U.S. newspaper, odds were high that the words, *National Wildlife Refuge* followed. The question of whether oil exploration and maybe drilling should be allowed in a part of the Refuge on the Arctic Coast has simmered for years.

The oil industry argues that the United States needs the oil to lower reliance on foreign oil, and that drilling would take place on less than one percent of the refuge land. Opponents argue that Americans should reduce oil consumption instead of drilling in a wilderness area.

Location of the Arctic National Wildlife Refuge.

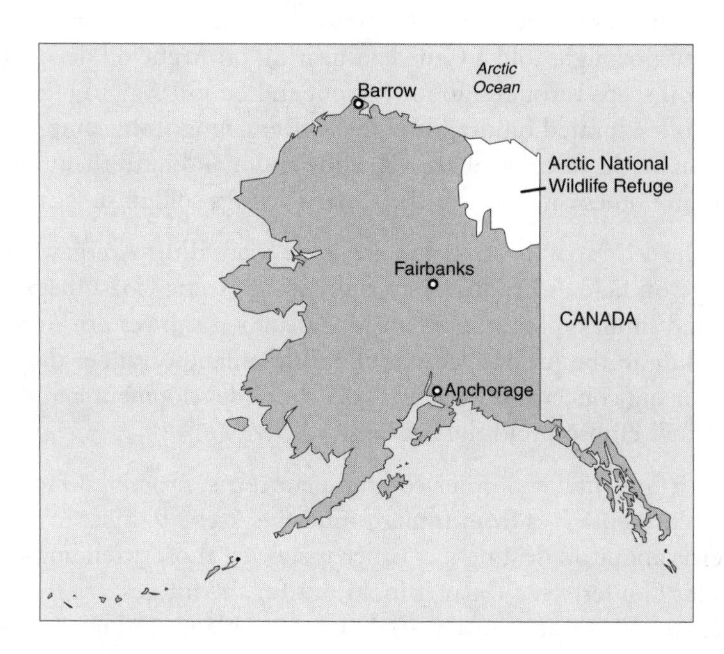

The Inupiat Eskimos who live along the coast generally favor opening the refuge to drilling because they feel operations at Prudhoe Bay have shown it does not harm wildlife. With tax revenue from the Prudhoe Bay oil complex falling, North Slope Borough officials look forward to the potential revenue from a new oil complex. But the Gich'in Athabascan Indians who live in northeastern Alaska, and northwestern Canada fear the project would disrupt the Porcupine Caribou herd's calving grounds. The caribou are an important food source for these Indians.

Although the Inupiat who live along the Arctic Coast generally favor drilling for oil on land at the Refuge, they are strongly opposed to any offshore drilling because of fears that oil spills are bound to occur and these could kill or chase away the whales that are such an important part of their lives. Sea ice is a danger because the keels of multi-year ice can cut into the ocean bottom in shallow water, ripping into pipelines.

"I don't know anyone here who want to see offshore development," says Glenn Sheehan of the Barrow Arctic Science Consortium. "Industry and government have demonstrated that they cannot even physically get in a position to respond to a spill in loose ice conditions. There are no widely accepted studies on what a spill might do to whales and other creatures."

Traditional Doesn't Mean Unchanging

Sometimes visitors to the Arctic are disappointed not to find Eskimos living the way they did before the first Europeans and Americans came into their lives. But look at it this way: We can be sure that an Inuit 500 years ago who devised an improvement to his harpoon wasn't chided for turning his back on tradition.

Glenn Sheehan, who studies Inuit culture, says: "My belief is that Eskimo tradition is to use the best of what's available, the proviso being that it help, or at least not interfere with the core value of whaling."

In Gretel Ehrlich's book *This Cold Heaven: Seven Seasons in Greenland*, we meet a Danish man who came to Greenland while young and learned to hunt and, to a great extent, live a traditional Inuit life. At the end of the book, he has moved to a larger town in Greenland to give his children a chance for the better education that will give them "a wider angle to make decisions later in life." His decision is like the one that parents in traditional cultures around the world face: Do we give our children a chance to choose the life they should lead? Or do we shelter them from the modern world with the hope they will follow the old ways?

The Dane in Ehrlich's book articulates a danger of the new ways: "It is easier to be a loser now … That was not possible before. Failure meant certain death."

He sees value in his children having led the old style life, even if they choose the new. The problems he hears people in Denmark talking about are really small. "I know how lucky the children and I are. These things don't bother us because we have lived with something bigger—we have lived, you could say, inside the weather, and out on the ice, being hungry and seeing people and animals die, and every day, just being there in the midst of beauty."

The Least You Need to Know

- For many people in the Arctic, life is a mixture of the traditional and the modern.
- In 1971, the United States settled native claims in Alaska with land and payments to native corporations.
- The Eskimo native corporation has done extremely well.
- Greenland is part of the Danish Kingdom with "home rule," but depends on aid from Denmark.
- Inuit from Alaska are helping Russian Inuit return to some traditional ways.
- Although Alaska's Eskimos approve of drilling in the Arctic National Wildlife Refuge, they oppose offshore drilling.

Chapter 16

No One Owns Any of Antarctica

In This Chapter

- Antarctica's future isn't what it used to be
- No one owns Antarctica, although some claim to
- The Cold War cools down in Antarctica
- A treaty makes Antarctica a continent for science
- Antarctica is no longer a boy's club

When you first glimpse Antarctica's vast expanses of ice with no trees, fields, or real cities you might think, *Surely the world's problems haven't followed me here*. That's not the case. Since early in the nineteenth century, villains as well as heroes have gone to Antarctica—and even the heroes were men of their times and cultures. Although some of the world's problems have migrated to Antarctica, the continent is different from the rest; it has never seen war. It also differs from the rest of the world because it has no native people and countries. No one really lives there; everyone is a visitor. In this chapter, we see how this unusual state of affairs came about, and what it's like today.

Countries Contended for Parts of Antarctica

Disputes over ownership of Southern Ocean islands, and even parts of Antarctica itself, go back to the explorers of the seventeenth century and disputed claims of who first saw or landed on certain islands. In those days, most European nations considered landing on an island, or a continent, good enough to claim ownership. Land-claim disputes simmered but didn't break out into war until 1982 when Argentina invaded the British Falklands Islands colony and the British recaptured it.

Between 1917 and 1933, Great Britain and France made claims based on exploration to large, pie-shape areas of Antarctica that met at the South Pole. The British claims were shared with the former colonies of New Zealand and Australia. In 1938, Australia, France, Great Britain, and New Zealand recognized each other's claims.

Explorations

In 1938, Hitler's Germany sent a seaplane "mother ship" and two seaplanes to establish an Antarctic claim. In January and February 1939, the planes flew over the part of Antarctica south of the Atlantic Ocean, taking photos of 96,500 square miles (250,000 square kilometers) of Antarctica. The airplanes also dropped 5-foot (1.5-meter) long, aluminum darts with swastikas engraved on the tops. The German claim prompted Norway to make a formal claim based on exploration to the pie-shape area between 20 degrees west and 45 degrees east longitude, extending almost to the South Pole.

The United States Considers a Claim

During his expeditions between 1928 and 1935, Richard Byrd "claimed" various parts of Antarctica for the United States, but the U.S. government's policy then—and now—is not to make any claims in Antarctica or to recognize the any other claims. However, in 1939, the U.S. State Department recommended that the United States claim land explored by Byrd plus a large part of the Peninsula. In 1939, when Byrd led the first U.S. government expedition to Antarctica in 100 years, his orders included:

> Members of the (expedition) may take any appropriate steps such as dropping written claims from airplane, depositing such writings in cairns, et cetera, which might assist in supporting a sovereignty claim by the United States government.

The possibility of U.S. claims prompted Chile to claim, in 1940, a pie-shape area that includes the Peninsula, and overlapped about half the British claim in that part of Antarctica. Chile's claim was based on the 1494 Treaty of Tordessillas, which divided the New World between Spain and Portugal. In 1943, Argentina made a claim for another pie-shape area overlapping most of the British claim and a large part of the Chilean claim in the Peninsula area.

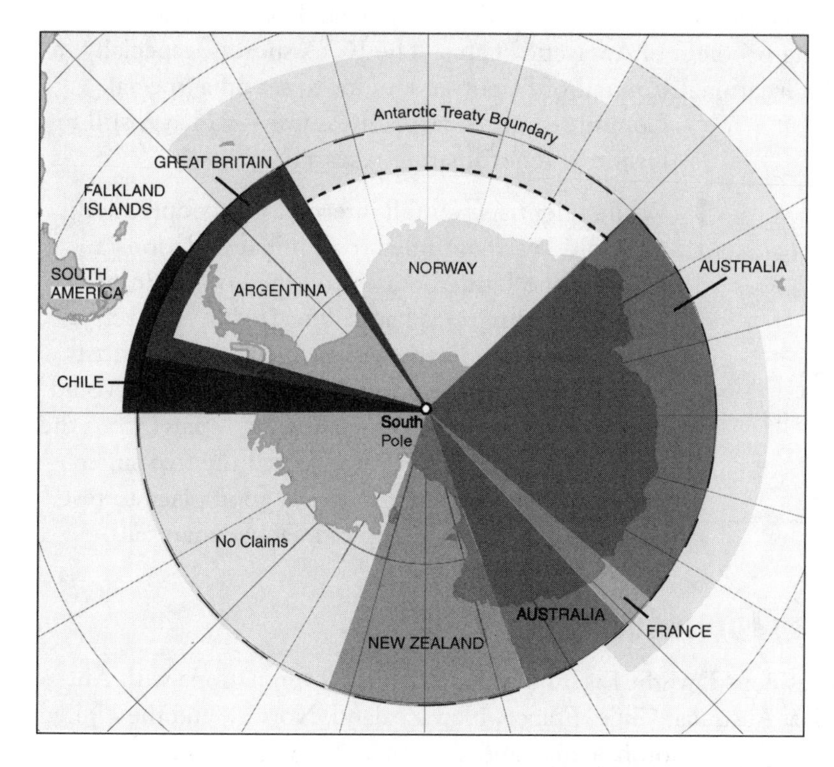

National claims to Antarctica.

(Information from Central Intelligence Agency, Polar Regions Atlas, *1978)*

Cold War Complicates Conflict Over Claims

By the 1950s, "The political situation in the Antarctic had grown chaotic, as overlapping national claims had carved the continent into an inedible pie with all slices meeting at the geographic South Pole," Paul Siple writes in *90° South*. "Oddly enough, little if any of the territory had even been seen by the vociferous claimants." By the 1950s, the United States had explored more of Antarctica than all other countries together, yet the United States was making no claims.

Byrd, Siple, and other American Antarctic explorers feared that nations with claims would squeeze the United States out of Antarctica. Of course, Operation Highjump in 1946 showed that the United States could use force to get its way, but no one wanted to start a war over Antarctica with British Commonwealth nations, France, Norway, Chile, or Argentina. Instead, the U.S. State Department asked Byrd, Siple, and others to officially document their explorations since 1928 to give the United States solid ground to make its own claims.

The Cold War Cools Off in Antarctica

As we saw in Chapter 14, the International Geophysical Year (IGY) was a great scientific success with nations coordinating their work and sharing data. The IGY also offered an answer to the vexing problem of Antarctic claims. The IGY's success, especially in Antarctica, led the International Council of Scientific Unions to set up a Special Committee on Antarctic Research—which is still in business—coordinating research.

Ice Chips

Before and during the IGY, Argentina, Chile, and the United Kingdom tried to reinforce their claims by setting up stations on the Antarctic Peninsula and nearby islands. For the IGY, these nations had 23 stations on the Peninsula, almost as many as on the rest of the continent.

While scientists wanted international cooperation, the Cold War continued to dominate relations between the United States and the Soviet Union. Still, the two nations had cooperated in Antarctica during the IGY. Each even allowed a few scientists from the other nation to work at its stations. Would these two nations continue allowing "spies" from the other into its stations? Some feared the two superpowers might see Antarctica as a good place to test nuclear weapons or dump radioactive materials.

A Treaty Sets Antarctica's Future Course

In late 1958, U.S. President Dwight Eisenhower invited the seven nations with Antarctic claims: Argentina, Australia, Chile, France, New Zealand, Norway, and the United Kingdom plus Belgium, Japan, South Africa, and the Soviet Union, which, like the United States, had IGY Antarctic stations, to negotiate the future of Antarctica. To the surprise of many, representatives from all the nations, meeting in Washington, D.C., signed the Antarctic Treaty on December 1, 1959. By June 23, 1961, all 12 nations had ratified it, putting the treaty into force.

Each nation agreed to inform all the others of its Antarctic plans. In addition, the treaty reads, "… all stations, installations and equipment (in the Antarctic) and all ships and aircraft at points of discharging or embarking cargoes or personnel in Antarctica shall be open at all times to inspection." This is important because the treaty says Antarctica is to be used for peaceful purposes only. Military forces can be used to support science, but no nation can establish military bases or fortifications, conduct military maneuvers, or test weapons.

The Treaty Sidesteps National Claims

Instead of trying to resolve the claims issue, treaty negotiators decided to put the claims on hold for the life of the treaty. No new claims are allowed, but existing claims were neither explicitly recognized nor denied. U.S. State Department officials credited the treaty with keeping the 1982 Falklands War from spreading to Antarctica. Even during the war, Argentina and Great Britain continued to cooperate in Antarctica. The treaty doesn't cover the Falkland Islands, but does include the ocean from 60 degrees south latitude south across the continent.

In 1997, the U.S. Antarctic Program External Panel, headed by Norman R. Augustine, then chairman and chief executive officer of the Lockheed Martin Corp, cited the scientific benefits of Antarctic stations, and also said: "The substantial U.S. presence in Antarctica is ... a critical, perhaps the most critical element in assuring the region's continued political stability."

Ice Chips

Nations with Antarctic claims talk as though they "own" parts of Antarctica. An Australian Antarctic Division press release says, "$102 million to Protect Australian Antarctic Territory." New Zealand considers the area where the U.S. McMurdo Station sits as being in its "Ross Sea Dependency." Unlike at home, however, New Zealanders in Antarctica drive on the right side of the road, like Americans, between McMurdo, New Zealand's Scott Base, and the nearby (U.S.) airfields.

It's Against U.S. Law to Violate Treaty Provisions

The 12 original nations plus another 15 which have "substantial scientific activity" in Antarctica and have signed the treaty, meet each year to discusses additions or changes. For a provision to go into force, all must agree to it. The treaty organization has no way to enforce the provisions; instead, it is up to each nation to enforce the treaty where it operates in Antarctica.

Crevasse Caution

When you're in Antarctica, don't pick up penguins. This violates the U.S. law against "harassing" wildlife. An example of how seriously this is taken: On January 19, 1999, a penguin on the runway at McMurdo delayed for about 10 minutes the takeoff of the airplane that carried the writer of this book to the South Pole. If the penguin hadn't moved, the plane would have waited for a researcher with a penguin-handling permit to arrive and move it.

The Antarctic Has a Long History of Exploitation

During the nineteenth and twentieth centuries, Europeans nearly wiped out seals and drove some species of whales close to extinction. Most commercial whaling near Antarctica ended in 1986, but disputes over Southern Ocean life continue.

In February 2002, chefs from chic and trendy restaurants in the San Francisco area announced they would no longer serve "Chilean sea bass" because the species was being depleted and most of the fish they had been serving was probably caught illegally. The fish, really the "Patagonian toothfish," became fashionable in the early 1990s when someone came up with the "Chilean sea bass" name.

Environmentalists and biologists are concerned because toothfish take a long time to mature, around 10 years before they are ready to begin making little toothfish. Also, toothfish are caught with lines strung out as much as 80 miles (130 kilometers) behind fishing boats with thousands of baited hooks. An albatross, spotting what looks like an easy snack, swoops down and finds itself hooked.

Ice Chips

Toothfish are a hot issue in Australia and New Zealand. On September 30, 2002, the Australian Broadcasting Corp. reported: "Toothfish have become a magnet to organized crime groups in search of a fast profit." Some Australian fishermen talk of putting armed men on their vessels to make "citizens arrests" of illegal toothfish boats, and the Australian and New Zealand navies chase and capture a few of the boats, which are seized along with their catches.

Since 1961, the Antarctic Treaty nations have struggled to come up with environmental regulations, balancing competing desires for development, making a buck, or peso, or pound, and preserving a unique environment. Environmental groups celebrated in 1997 when treaty nations ratified a far-reaching environmental protection protocol. Among other things, it bans mining in Antarctica for 50 years and designates the continent and its marine ecosystems as a "natural reserve devoted to peace and science."

Stations Begin Taking Out the Trash

When the U.S. Navy first began developing the McMurdo Station during Operation Deepfreeze in 1955, the least of its concerns was keeping the place pristine. The men at McMurdo and other stations tossed trash and garbage into big, open dumps; no one worried when oil or gasoline spilled onto the ice or rocky soil; sewage of all kinds flowed into bays and the ocean. Stations around the Antarctic Coast often left trash on the sea ice to fall into the ocean when the ice melted. The mess was the biggest at McMurdo because it's the biggest station, but no Antarctic station would have qualified for a Good Housekeeping award before the 1990s.

One of the first things a newcomer at McMurdo notices today are the rows of recycling bins outside many of the buildings; the most obvious indications of the major clean up that began in 1988. Today all trash and even kitchen and dining room garbage is shipped back to the United States for recycling or disposal.

Stations around the coast that have been dumping sewage into the ocean are building treatment plants. At inland stations, such as the South Pole, sewage flows into a huge pit in the ice, which means a very dirty iceberg is likely to break off the Ross Ice Sheet around 150,000 years from now—ice from the center of the continent flows very slowly to the sea.

Antarctica Is No Longer Where Only the Boys Are

The first woman to step foot on the Antarctic continent was Caroline Mikkelsen, the wife of Llarius Mikkelsen, the caption of a Norwegian whaling fleet ship, the *Thorshaven*. On February 20, 1935, Captain Mikkelsen, his wife, and seven crew members set out in a small boat to briefly explore the part of the Antarctic Coast, now known as the Vestfold Hills. They spent about six hours ashore, putting up a Norwegian flag, building a small cairn, and naming the area "Ingrid Christensen Land" in honor of the wife of Mikkelsen's boss, Lars Christensen. (That name didn't stick.)

The road from Caroline Mikkelsen's brief visit to women running Antarctic stations, as they do today, was anything but easy.

An example: In December 1955, Richard Byrd and Paul Siple changed planes in Dallas, Texas, on their way to New Zealand to go on to Antarctica. Byrd "was highly amused by girl pickets, aspiring to explore Antarctica," Siple wrote. The women's signs said: "Byrd unfair to women," because he had been quoted as saying "Little America was the quietest place on Earth because no woman had ever set foot there." Byrd's attitude was the norm into the 1960s.

In 1946, Edith Ronne planned to travel by ship as far as Valparaiso, Chile, with her husband, Finn, who was on the way to spend the winter on the Peninsula as leader of a private expedition with 21 men. But when they arrived in Chile, Finn persuaded her to continue to Antarctica to file the newspaper reports he had contracted to send. The wife of Harry Darlington, one of the pilots, wanted to go along also, and talked her husband and Finn Ronne into it. Despite the traditional fears that women would disrupt an expedition, especially one spending the winter, the women's presence didn't cause any problems. But that lesson wasn't taken to heart for the next quarter-century.

The Navy Ends Its Policy of Men Only on Planes

Until 1969, the U.S. Navy wouldn't allow women to board its airplanes going to Antarctica, and the National Science Foundation went along. But things were changing. For the 1969–1970 season, the Navy said women could fly to Antarctica.

By then, the National Science Foundation had encouraged a few women scientists to apply for Antarctic projects. The first woman to take part in the U.S. Antarctic program was Christine Muller-Schwarze, who with her husband, Dietland, studied the Cape Crozier penguin colony. They reached McMurdo early in the season, in mid-October 1969, to be on hand when the penguins arrived.

Ice Chips

The women who went to Antarctica in 1969 encountered an unexpected danger: the press. In *The New Explorers: Women in Antarctica,* Barbara Land gives an example of questions they got: "Will you wear lipstick when you work?" They saw headlines like: "Powderpuff Explorers to Invade South Pole." Walter Sullivan of *The New York Times* was an exception, Lois Jones told Land. "He seemed really interested in our work and explained what we were trying to do."

Also working in Antarctica in the 1969–1970 season were Lois Jones, an Ohio State University geochemist, leader of a four-woman team that included Terry Lee Tickhill, Eileen McSaveney, and Key Lindsay. They lived in tents in the Dry Valleys. Soon after they arrived at McMurdo, Jean Person of the *Detroit Free Press,* the first woman journalist in Antarctica, showed up. The seventh woman in Antarctica that year was Pam Young, a New Zealand biologist at the Scott Base, near McMurdo.

The Navy invited the women to make a quick trip to the South Pole in November 1969. Christine Muller-Schwarze turned down the trip; she was too busy with the penguins. Rear Admiral Kelly Welch, that year's Naval commander in Antarctica, arranged for the LC-130 they were on to lower the rear ramp when it pulled up at the pole station. The six women then linked arms—with the admiral in the middle—walked down the ramp, and stepped onto the ice together. Each of the six was the "first woman at the South Pole."

In the beginning, the women who went to Antarctica with the U.S. program were mostly scientists, but over the years more and more women have taken all kinds of jobs and today about one-third of the scientists and support staff are women.

(Jack Williams)

Students at a survival school on the Ross Ice Shelf in Antarctica take a break. On the left is a Scott tent, and on the right (behind the people) is a hollowed-out snow mound shelter.

(National Science Foundation, Melanie Conner)

A Long Duration Balloon carrying astronomical instruments is launched from the Ross Ice Shelf, Antarctica.

(Jack Williams)

Ice melting from the edge of the Greenland Ice Sheet, background left, feeds the Watson River near Kangerlussuaq. Tourists often hike or ride to this area for picnics.

(Jack Williams)

The Kangerlussuaq Airport is Greenland's main air hub with regular jet service from Copenhagen, Denmark, and flights in smaller airplanes and helicopters to other parts of the island.

(Jack Williams)

The Kangerlussuaq Tourism Office's four-wheel-drive vehicles take visitors to the edge of Greenland's Ice Sheet for a picnic.

(Jack Williams)

Greenland's Jacobshaven Glacier as seen from a low-flying airplane. As the ice melts, dust and sand blown onto the ice sheet over thousands of years remain.

(National Science Foundation, Melanie Conner)

Tourists from the cruise ship Kapitan Khlebnikov *gather outside the hut that Robert Scott built in 1902 during his first Antarctic expedition. It's on the edge of today's U.S. McMurdo Station.*

(National Science Foundation, Daniel Dixon)

The U.S. International Trans-Antarctic Scientific Expedition team at the ceremonial South Pole in January 2003. Paul Mayewski, the expedition leader, is in yellow coveralls at left. The new South Pole Station is behind them.

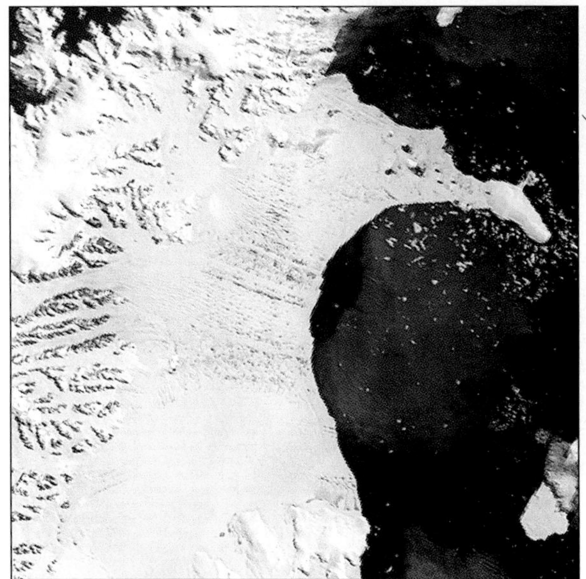

(NASA Terra satellite image from the National Snow and Ice Data Center)

The Larsen B Ice Shelf on the Antarctic Peninsula on January 31, 2002. The blue spots on the ice shelf are ponds of water from melted ice. The dark area on the right is the Weddell Sea. The Peninsula's mountains and glaciers are on the left. The image shows an area 101 miles (162.5 kilometers) across.

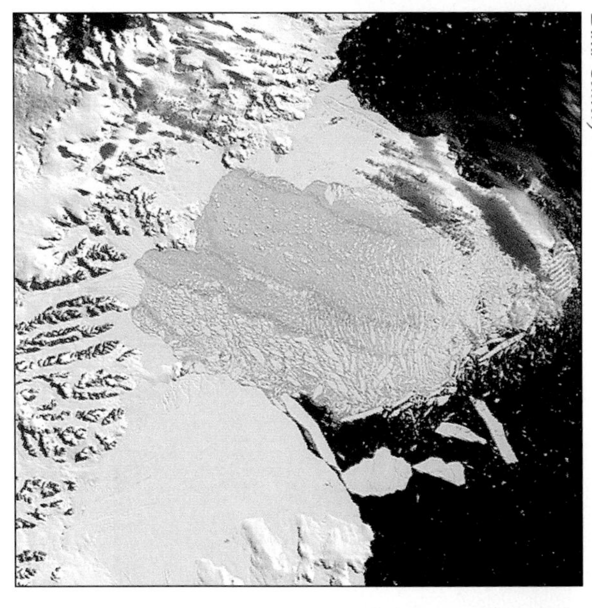

(NASA Terra satellite image from the National Snow and Ice Data Center)

The same scene as in the photo above, on March 7, 2002, after the ice shelf collapsed (discussed in Chapter 27). Pieces of the ice shelf that have turned on their sides are azure blue. Pieces that are still upright are white. Large pieces of the shelf are floating away as icebergs at the bottom right. The flat, white area at the bottom is the part of the ice shelf that did not break up.

(National Science Foundation, Melanie Conner)

The new main building for the U.S. Amundsen-Scott South Pole Station under construction in December 2002.

Interior of the dome at the South Pole Station. Inside the dome is as cold as outside, but there is no wind. Men and women at the pole live and work in the red buildings.

(National Science Foundation, Melanie Conner)

(National Science Foundation, Melanie Conner)

A skua tries to chase an Adelie penguin away from its nest so the skua can eat the egg the penguin is incubating. They are in the Cape Royds, Antarctica, penguin rookery.

(National Science Foundation, Josh Landis)

The northern edge of the giant B-15A iceberg in the Ross Sea, Antarctica, on January 29, 2001.

(National Science Foundation, Daniel Dixon)

The U.S. International Trans-Antarctic Expedition crossing the ice in December 2002. The tractor at right is one of two that pulled trains of supplies and the tents and huts members lived in. The "polar pooper" brings up the rear. At each stop a hole was dug and the portable outhouse pushed over it.

(National Science Foundation, Peter West)

The U.S. Antarctic Program field camp at Lake Hoare in the McMurdo Dry Valleys. The Canada Glacier is in the background.

The Least You Need to Know

- Since early in the twentieth century, seven countries have claimed to own parts of Antarctica.
- The United States has never claimed any part of Antarctica and has never recognized any Antarctic claims.
- In 1961, the nations active in Antarctica ratified the Antarctic Treaty, which forbids military activity in Antarctica.
- A protocol to the Treaty ratified in 1997 bans mining in Antarctica for 50 years.
- U.S. stations in Antarctica send all the garbage to the United States for disposal or recycling.
- Women were first allowed at U.S. Antarctic stations in 1969, and today about one third of the Americans in Antarctica are women.

Chapter 17

Taking to the Air in Antarctica

In This Chapter

- The Navy makes airplanes routine transportation
- Hercules begins doing the heavy lifting
- An airplane flies after 16 years under the snow
- Air National Guard takes over Antarctic flying
- Airplanes rescue two ill South Pole physicians

Although polar explorers began using airplanes in the 1920s, they still relied on ships to reach the polar regions, and dog teams and later snowmobiles and tractors to get around. Although this is still true for many who go to Antarctica for other nations, the United States uses mostly airplanes both to travel to and from Antarctica and to get around the continent. In this chapter, we'll see how the U.S. Navy pioneered Antarctic aviation in the 1950s and then look at the challenges that today's polar aviators face.

Airplanes Become More Than Pathfinders

As we saw in Chapter 14, the U.S. Navy began making regular airplane trips to and from Antarctica and around the continent in the mid-1950s. Navy and U.S. Air Force airplanes hauled everything needed to build and supply the first South Pole Station in 1956, and during the 2003–2004 season New York Air National Guard airplanes will continue hauling in materials for the third South Pole station, now under construction. When it's finished, everything for the three stations the United States has built at the Pole will have arrived by airplane.

We also saw in Chapter 14 how the first airplane that arrived to bring home the 18 men who had spent the first winter at the South Pole, in October 1957, ended up stranded there with a failed engine. Such breakdowns were common with the *piston-engine* airplanes the Navy and Air Force flew in the 1950s.

Polar Talk

A **piston engine** uses the expansion of a burning gasoline and air mixture inside cylinders to push pistons down to turn a crankshaft. The crankshaft turns the propeller.

Airplane piston engines are complex and produce a lot of vibration, which means problems such as broken fuel lines can occur all too often. These engines failed and even caught fire in flight more often than an airline or the military today would tolerate. This is why almost all of today's airline and military airplanes use *turbojet* and *turboprop* engines.

Polar Talk

A **turbojet** (jet) takes air in at the front. The air is mixed with fuel and burns, sending a powerful jet exhaust out the rear, which propels the airplane. On the way out, the exhaust spins blades attached to a shift, called a turbine, which supplies power for devices such as the engine's fuel pumps. Jets work best at high speeds and altitudes.

A **turboprop** is an aircraft engine that uses the turbine shaft to turn a propeller. It combines the smooth efficiency of a turbine with a propeller's efficiency at low speeds and altitudes.

In the mid-1950s, to counter the Cold War threat of an attack by Soviet bombers coming across the Arctic Ocean, the United States built the Distant Early Warning (DEW) line of radar stations stretching 3,000 miles (4,830 kilometers) from northwestern Alaska to Baffin Island, Canada.

When the Air Force expanded the DEW line with two stations on Greenland's Ice Sheet, it needed a way to haul in construction materials and supplies. The solution: Add skis that fit around the landing gear of a C-130 Hercules transport airplane. To take off and land at an airport, the skis lift up and the wheels come down through slots in the skis. To land on ice, the skis are lowered below the wheels. Adding skis makes the Hercules an LC-130.

In addition to more reliable turboprop engines, the LC-130 has other advantages over other ski-equipped military cargo airplanes. An LC-130's large rear ramp allows snowmobiles and tractors to drive on and off the airplane. The Air Force begin using LC-130s in Greenland in the late 1950s, and the Navy begin flying LC-130s in Antarctica in 1961.

Airplane Spends Fifteen Years Buried in the Snow

Switching to LC-130s made polar flying more efficient and safer, but danger still abounds. For instance, on December 4, 1971, a Navy LC-130 delivered supplies to a French *traverse* party about 850 miles west of the McMurdo Station. The pilot was using assisted takeoff rockets, which the Navy called JATO for "Jet Assisted Takeoff," when about 50 feet above the ice, two of the 165-pound rockets broke loose and hit a propeller, ripping it off, damaging another engine and sending debris flying into the fuselage. The pilots managed to land and no one was injured. An investigation team ruled the airplane a total loss and left it on the ice to be covered by snow over the years.

Polar Talk

Among polar scientists, **traverse** refers to research using a party traveling across ice by snowmobile, or tractors pulling sleds with equipment and shelter. The party stops along the way to make observations.

In the mid-1980s, the National Science Foundation needed another airplane and a study concluded that recovering the "ice plane" would cost maybe one-third the $35 million price of a new LC-130. In November 1986, a French and American team, using tractors, pulled equipment to the crash site, where the only sign of the LC-130 was the top 3 feet (1 meter) of the tail sticking out of the snow. That year the team smoothed out a skiway for LC-130s to use, dug the airplane from the snow, and determined it could be repaired. The following year, the recovery team installed new engines and flew the LC-130 back to McMurdo. After repairs in the United States, the Navy used it in Antarctica until February 1999.

Ice Chips

Even today, jets aren't used for flying within Antarctica, but Air Force C-5 and C-141 jet transports fly from Christchurch to land on an ice runway near the McMurdo Station. In addition to being faster than the LC-130s, they carry more people. Using them allows the LC-130s to spend their time flying from the groomed snow skiway at McMurdo to skiways at the South Pole and field camps where airplanes couldn't land or take off without skis.

The Air National Guard Takes Over in Antarctica

The Navy's long tradition Antarctic flying ended on February 17, 1999, when the last Navy LC-130 took off for Christchurch, New Zealand, from Williams Field near the McMurdo Station. Even though the National Science Foundation was paying the Navy to fly in Antarctica, "the Navy reluctantly reevaluated its involvement in Antarctica and decided the resources being used should go to supporting carrier battle groups," Commander David W. Jackson, the last Navy commanding officer in Antarctica, said.

A cloud-cloaked Mount Erebus looms over LC-130s on the ice ramp near the McMurdo Station. All maintenance work is done outside; there are no hangers.

(Courtesy the National Science Foundation, Josh Landis)

The New York Air National Guard's 109th Airlift Wing, which had been helping the Navy in Antarctica for 11 years, took over the job, also using LC-130s. This unit's history goes back to the late 1950s Air Force squadrons that supported the DEW Line sites on the Greenland Ice Sheet. When these DEW line stations were closed, the National Science Foundation began using the 109th to support its Greenland research, and this led to Antarctica. Today, the 109th is the only military unit in the world capable of routinely landing large transports on ice—and taking off again.

Ice Chips

The 109th Airlift Wing is based at the Stratton Air Guard Base at the Schenectady County, New York, Airport. Pilots and crew train in Greenland, in addition to supporting science, during the Northern Hemisphere summer. Its Greenland base is the Kangerlussuaq International Airport, but flight crews learn and hone ice-flying skills at the Raven Skiway, a former DEW Line site on the Ice Sheet. It also conducts "Kool School," survival training there.

With the 109th Airlift Wing, the National Science Foundation has the aircraft and people needed to do polar heavy lifting in Antarctica and Greenland. If the U.S. military ever has a need to land troops on ice anywhere, the 109th has the airplanes to do the job. Even more important, with men and women who are experts in polar flying, the military wouldn't have to re-invent heavy-duty polar flying.

Polar Flying Is Never Routine

Polar flying, especially in Antarctica, is vastly different from flying anywhere else. Imagine flying from Dallas, Texas, to Minneapolis, Minnesota, without seeing anything below but ice, with maybe a bare mountaintop sticking out of the ice now and then. This is the distance of Antarctica's busiest air route, from McMurdo to the South Pole. Pilots have no en-route radar and no air traffic controllers keeping track of them. The only airports are the ones at each end of the McMurdo-Pole route, and both are skiways of groomed ice marked by black flags along the edges.

Landings and Takeoffs Are Often Tricky

Landing and taking off from a skiway is as close as Antarctic flying gets to routine. The biggest challenge is an "open field" landing and taking off from places without a groomed skyway. Before flying to such places, crews study satellite and aerial photos. When the airplane arrives, the pilots fly low over the site while the crew looks closely for any signs of crevasses.

Landing is just the beginning. If the snow is too soft, taking off can be impossible, especially at the 9,000-foot (2,740-meter) plus elevations of the Polar Plateau where the thin air robs airplanes of some of their engine power, propeller thrust, and lifting force. Once on Greenland's Ice Sheet in the 1970s, one of the 109th's airplanes was unable to take off in soft snow. The pilot started taxiing across the ice and trying to take off every once in a while. Finally, after taxiing 20 miles, the airplane had burned enough fuel to be light enough to get into the air.

Accidents in Antarctica, such as the one in 1971 that left an airplane abandoned on the Ice Sheet, prompted the Navy to stop using rockets for takeoffs. But over the years, the 109th Airlift Wing put a lot of effort into making rocket takeoffs safe, and ATO, the 109th's preferred name, are a normal part of its operations in Greenland and Antarctica.

A Crevasse Snags an Airplane

On November 16, 1998, the left, main ski of an Air National Guard LC-130 broke through an ice bridge and lodged in a 140-foot (43-meter) deep crevasse while it was taxiing to take off from a remote Antarctic field camp without a marked skiway. None of the 7 crew members and 14 passengers on the plane were injured, but the plane was stuck with unknown damage. A snow bridge about 8 feet (2.4 meters) thick had hidden the crevasse, which was about 10 feet (3 meters) wide.

As the first step toward freeing the airplane, engineers from the Army's Cold Regions Research and Engineering Lab in Hanover, New Hampshire, with ground-penetrating radar, along with mountaineers from the McMurdo Station, searched the area for a close-by, crevasse-free landing area and marked safe paths for those who recovered the airplane.

Then, technicians used plywood, snow, and 12-by-12 foot (3.6-by-3.6 meter) airbags and a lot of digging to level the airplane. This enabled a tractor to pull the plane out of the crevasse. Working in the cold, technicians replaced an engine and two propellers, and a crew flew the plane back to McMurdo on January 4, 1999. After a thorough inspection, the airplane was put back to work.

Weather Is the Biggest Challenge to Polar Pilots

In December 1997, two Navy LC-130s, one from Christchurch, New Zealand, the other from the South Pole, arrived at McMurdo without enough fuel to go somewhere else, just as a storm blew in, creating a *whiteout.* In a similar case, a pilot described landing in a whiteout as "like landing in a milk bottle." When this happens, the first choice is to fly somewhere with better weather. But in Antarctica the choices are limited to the South Pole and maybe a field camp with room for everyone on the plane to get inside out of the cold. Otherwise, the plane flies to an area of the Ross Ice Shelf that that's free of crevasses. Then the plane makes a whiteout landing, which both the Navy planes did in 1997.

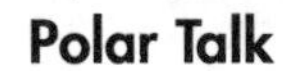

Polar Talk

A **whiteout** is a dangerous weather phenomenon that occurs when thick, low clouds reduce surface definition, and the horizon is obscured. Judging distances is difficult.

To perform a whiteout landing, the airplane's navigator ensures that the airplane will touch down in a designated area that's free of crevasses and other hazards. The pilots slowly descend, without being able to see the ice, until the airplane touches down. When this happens, says Colonel Graham Pritchard, who retired as commanding officer of the 109th Airlift Wing in 2000, "The pilot has to be very careful not to jerk the yoke back when you feel the skis touch down"—a natural reaction. If the pilot does pull back, the airplane will zoom up, and then probably come down hard. It could even bounce across the snow and crash.

After the December 1997 whiteout landing, members of the McMurdo Search and Rescue Team, in tracked vehicles equipped with global positioning satellite navigation systems and radar, quickly located the airplanes and began ferrying passengers to shelter. Even with the navigation system and radar, the last of the passengers didn't reach shelter until six hours after the airplane landed.

This is why passengers put on long underwear, polar fleece jackets and pants, wind pants, insulated boots, and parkas in 70-degree temperatures in Christchurch, New Zealand, before boarding an airplane for Antarctica.

Explorations

LC-130s are too big for some of the flying the National Science Foundation needs in Antarctica. For years the Navy supplied smaller airplanes as well as helicopters. Today, for jobs requiring smaller airplanes, the NSF contracts with Ken Borek Air of Canada, which flies Canadian-built de Havilland Twin Otters, an airplane designed to fly in the cold. These airplanes carry small science parties to field camps and are sometimes fitted with equipment for tasks such as radar mapping of the ice. The NSF also contracts with Petroleum Helicopters of Lafayette, Louisiana, for the helicopters that fly to the Dry Valleys among other jobs. The company started out flying workers, supplies, and equipment to Gulf of Mexico oil drilling rigs but has obviously branched out. Coast Guard icebreakers also bring helicopters to Antarctica.

Antarctica's Weather Is Hard to Forecast

All pilots need good weather forecasts, but Antarctic pilots especially need them because they have so few options when surprised by foul weather. Yet Antarctica's bitter cold and frigid winds that often roar downhill to a stormy ocean creates some of the world's most extreme weather. Making forecasts requires data on what the weather is doing now over a wide area, and this data is hard to come by in Antarctica.

Antarctica is larger than the contiguous 48 states; yet forecasters have a lot less data to work with. In the United States, hourly reports come from around 1,000 regular weather offices plus about 10,000 automated weather stations. Data flows in from the network of Doppler radars that watch almost all the United States. About 50 weather stations launch weather balloons twice a day to collect data from high above the earth. In Antarctica, data comes in from around 20 full-time weather stations, less than 200 automated stations, and maybe 10 places that launch weather balloons. There is no weather radar, and weather satellites don't look down on the polar regions 24 hours a day as elsewhere on earth.

The first step toward improving forecasts is collecting more data. So the National Science Foundation supports the system of automated Antarctic weather stations, which is now operated by the Space Science and Engineering Center, University of Wisconsin—Madison. The stations collect data and beam it up to satellites that transmit it to the University of Wisconsin, which makes it available.

Explorations

Data from Antarctica's automated weather station helped lead Susan Solomon, a National Oceanic and Atmospheric Administration scientist, to conclude that unusually bad weather on the Ross Ice Shelf in 1912 helped kill Robert Scott and the four men with him on their way back from the South Pole. She and Charles Sterns, who runs the automated stations program for the University of Wisconsin, used station data to show that weather such as what Scott ran into on his way back from the pole was unusually cold—the kind of cold that occurs on average only once in 17 years. Solomon says that even though Scott made many mistakes, "these alone would clearly not have been sufficient to cause all of them to die had the weather been normal." She tells the story of Scott and Antarctica's weather in *The Coldest March*.

Antarctica Doesn't Stump New Computer Model

Data alone won't create better forecasts for Antarctica. The National Center for Atmospheric Research in Boulder, Colorado, and the Polar Meteorology Group at

Ohio State University worked together to create the Antarctic *Mesoscale* Prediction System (AMPS), a computer model that combined features of other forecasting models with information about Antarctica's unique weather makers, such as the coming and going of sea ice.

Airplanes Rescue Two South Pole Doctors

The worst nightmare of anyone who signs up for the winter at the South Pole happened in early March 1999 when Dr. Jerri Nielsen, the only physician among the 41 men and women at the pole, discovered a lump in her right breast. In late June, she performed a biopsy on herself with the help of others at the station, and advice from physicians in the United States via an Internet video link. Unfortunately, images weren't clear enough for the physicians to tell whether she had cancer. On July 10, 1999, an Air Force C-141 Starlifter flew nonstop from Christchurch, New Zealand, and dropped by parachute boxes with drugs to treat cancer, equipment for sending clearer images back, mail, and some fresh fruit and vegetables, before returning to Christchurch without a stop. Those at the pole lighted oil-soaked wood in steel barrels as beacons for the airplane.

> **Polar Talk**
>
> In meteorology, **mesoscale** refers to phenomena that range in size from a few miles to around 60 miles (100 kilometers) across. These include thunderstorms, which do not occur in Antarctica, to local winds and small-scale storms, which are a big problem there.

Landing wasn't possible in the pole's –80°F (–62°C) temperatures. Machines needed to smooth the runway wouldn't run long enough to do the job, and the airplane's fuel and hydraulic fluid could have turned to jelly on the ground. (It's much colder at the ground at the South Pole in winter than higher up.) Finally on October 16, 1999, a New York Air National Guard LC-130 landed at the South Pole, another physician got off, Nielsen was helped aboard, and the airplane took off, taking her back to Christchurch, and eventually home. Nielsen recovered, worked with Maryanne Vollers to write a book, *Ice Bound: A Doctor's Incredible Battle for Survival,* and lectures about her experience. A CBS made-for-television movie starting Susan Sarandon, first seen in April 2003, also told her story. Then it happened again. The 2001 winter physician, Ron Shemenski, passed a gallstone in early April and had pancreatitis.

While he said he didn't need to be evacuated, physicians back in the United States said he should be brought back to the United States if he could. If his condition became worse, he could die.

This time the temperatures were on the border of being suitable for an Air National Guard LC-130 to make the trip, but the NSF decided to use Twin Otter airplanes flown by Ken Borek Air of Canada. The airplanes are slower, but simpler than the LC-130, and are designed to operate in colder temperatures. Two Twin Otter airplanes flew from Calgary, Canada, across the United States and South America to Punta Arenas, Chile. From there they flew to the British Antarctic Survey base at Rothera, on the Antarctic Peninsula.

Steve Dunbar, the field science support manager for Raytheon Polar Services Corp., coordinated the rescue from Punta Arenas using his knowledge of the people who fly in Antarctica and their aircraft. "I wouldn't have sent them if I hadn't thought there wasn't a 100 percent chance of their getting there," he said. If the plane had been forced to land in Antarctica, those aboard had a tent, sleeping bags for very cold weather, a week's supply of food, a stove, and shovels they could use dig snow caves or smooth a place for the other Twin Otter to land. If they had landed in a place where the other Twin Otter couldn't rescue them, Dunbar says, "We could have dropped supplies to them. We could have kept on dropping supplies as long as needed."

Weather would obviously be a key, and forecasters were using the new Antarctic Mesoscale Prediction System computer model, which had gone into use only the previous October. When forecasters said the weather should be good, one of the Twin Otters, piloted by Sean Loutitt and Mark Cary, and with flight engineer Peter Brown, took off for the South Pole. With them was Betty Carlisle, who would replace Shemenski as the pole physician. After a nine and a half hour flight to the pole, the crew rested 12 hours while heaters kept the airplane warm. It then safely returned to the Rothera Station and Chile with Shemenski.

Loutitt, the pilot, said the flight to and from the pole in the dark, at a time of the year when no one usually flies in Antarctica, wasn't as challenging as his summer flying there. Temperatures might be 50° to 75°F (28° to 42°C) warmer in the summer, but, "On this flight we had a lot of good ground support. During the summer, we go into places no one has ever been to before. The onus is on you and the flight crew to use judgment."

The Least You Need to Know

- Beginning in the 1950s, the U.S. Navy made airplanes the main form of Antarctic transportation.
- The turboprop LC-130 "Hercules" transport airplane, which went into service in the late 1950s, is still used.
- Jets fly to and from Antarctica, but not around the continent.
- The New York Air National Guard took over U.S. Antarctic flying in 1999.
- Providing the weather forecasts Antarctic pilots need is extremely difficult.
- Daring aerial rescues of ill physicians at the South Pole illustrate how vital aircraft are in Antarctica.

Chapter 18

Going to Work in Antarctica

In This Chapter

- Getting to Antarctica is easier than it used to be
- Antarctic workers feel an affinity with its history
- Life at the McMurdo Station today
- Today's technology doesn't make Antarctica safe
- Building a new South Pole Station
- How to get an Antarctic job

Since at least the middle of the nineteenth century, travelers to Antarctica have called it "The Ice," and ice defines Antarctica as much now as it did then. Many of today's Antarctic visitors are fascinated with Heroic Age explorers because the icy environment is the same. In this chapter, we'll look at how Antarctica has changed over a century. Today's researchers are more comfortable than those of in the past, but like them, most appreciate being among the few people ever to live on The Ice.

Antarctica Is Still a Long Way Away

If you go to Antarctica with the U.S. Antarctic Program, you are likely to begin your journey at the International Antarctic Centre in Christchurch, New Zealand. Here you pick up two canvas bags with the "extreme cold weather" clothing you'll need, and ensure everything fits because tight clothing could reduce your blood's circulation, making frostbite more likely.

To get to Antarctica, you might board a U.S. Air Force C-141, which jets you the 2,500 miles to the McMurdo Station in around 5 hours. Or you might go in one of the New York Air National Guard LC-130 "Hercules" turboprops, which takes eight hours. If the weather is acting up, you might spend days in Christchurch—a nice place to wait. If you are unlucky, you will board an LC-130, fly for eight hours, and land back in Christchurch. You have *boomeranged.*

Polar Talk

A flight that lands at the airport it took off from is a **boomerang.** This usually occurs in Antarctica because of an unforecast weather change before the airplane reaches the "point of safe return," when it has enough fuel to return.

Antarctica's Weather Can Still Kill You

Heroic Age explorers kept an eye on their companions, to tell them when their cheeks or noses were beginning to turn white—the first sign of frostbite. This is true today, although today's Antarctic travelers must make sure their warm dormitory rooms don't lull them into being careless about Antarctica's dangers when they venture outdoors.

Those living at McMurdo today have hot food for breakfast, lunch, dinner, and *midrats* in a comfortable, warm dining room. You have a choice of fresh fruits or vegetables flown in from New Zealand. At the small store you can buy snacks, a music CD, or a bottle of whiskey as a gift to a camp you will visit. No cash? Use the ATM.

But if a storm blows in with strong winds, bitter temperatures, and zero visibility, you will be stuck in your dormitory, the dining hall, or maybe the Albert P. Crary Science and Engineering Center, because when conditions are bad enough, no one is allowed to go outside. The science center would be at home on any university campus in the United States. If you are trapped there you could hang out in the library and computer center, chatting via e-mail with friends back home. When you can see the breathtaking view from the library of the Transantarctic Mountains around 43 miles (70 kilometers) away across McMurdo Sound, the storm is over.

Polar Talk

Midrats is a military term for "midnight rations" for those who work at night. Often the rule is that night workers have priority over those who just want a midnight snack.

Ice Chips

Heroic Age explorers could have only dreamed of the food available today in Antarctica. One day, the waffle stand in the McMurdo galley had a sign: "Due to the fact that it is a harsh continent, we are out of strawberries, management regrets any inconvenience."

Living in the Field Isn't Quite as Comfortable

While some scientists work at McMurdo, or close enough to live there and commute by snowmobile or helicopter to their research sites, for many, McMurdo is a stop on the way to and from the South Pole Station, a "field" camp, or a "deep field" location. A field camp is one with semi-permanent buildings, usually Jamesway huts, open only during the summer. One of the Jamesways is probably the galley and there is a full-time cook or two, depending on the camp's size.

The Siple Dome field camp in 1999. The Jamesway hut on the left is the workshop for the ice-coring drill tower at the far end. Two outhouse toilets, the only kind at the camp, are on the right.

(Jack Williams)

At field camps, and smaller stations such as the South Pole, everyone pitches in to keep the place running. Each person will be the *housemouse* one day each two or three weeks. This generally means helping the cook by washing dishes and cleaning the kitchen, maybe doing other cleaning.

At many camps, including even the South Pole where temperatures rarely are as warm as 0°F (–18°C), many people prefer to sleep in their own tent rather than a Jamesway where at best your "room" has hanging blankets for walls, and you hear others snore.

Deep field camps are smaller and everyone is likely to live in tents, probably with a large tent serving as a galley and general gathering place. Sometimes researchers might live for a few weeks in such a camp, either in the same place or traveling on snowmobiles pulling sleds loaded with supplies and equipment.

Polar Talk

The person at a polar research camp who spends a day doing various domestic chores, such as cleaning the kitchen, is the **housemouse.** This duty rotates among everyone in the camp.

Polar Talk

Reverse osmosis is a method for removing salt from seawater. In normal osmosis, a less concentrated liquid, such as fresh water, will flow through a membrane into a more concentrated liquid, such as salt water. Applying enough pressure to the salt water will force it through a membrane, filtering out salt.

Those who work in Antarctica have one thing in common with astronauts: people want to know, "how do you go to the bathroom?" At permanent stations, such as McMurdo, you enjoy heated bathrooms with flush toilets, and running hot water for showers. McMurdo discourages lengthy showers because the station uses relatively costly *reverse osmosis* to make fresh water from salty seawater.

At the South Pole, hot water is run into a well that goes about 350 feet (106 meters) into the ice. The hot water melts ice, which is pumped up and to different parts of the station after being filtered. At smaller camps, which have generators, someone scoops up snow with a front-end loader and dumps it into a bin that's heated by the exhaust pipe from the generator's engine.

At field- and deep-field camps on the ice, the toilet is a black outhouse where the waste drops into a hole in the ice. Even in frigid temperatures these aren't as cold as you might think, although no one would call them warm. With the sun up 24 hours a day, the tiny black buildings absorb heat. Also their Styrofoam toilet seats aren't as cold as plastic would be.

A New Station Is Going Up at the South Pole

The South Pole is a very special place to live and work. Almost everyone who goes there realizes that the station is in Earth's harshest permanently occupied environment. Not only does the temperature rarely climb to 0°F (–18°C), its 9,200 foot (2,800 meter) elevation means *altitude sickness* is a constant danger.

The experience during a South Pole visit by the author of this book is typical: Wake up at 2 A.M., thirsty, trying to suck in air my body is demanding. It's dark and warm inside the Jamesway, but bright light streams through a tiny opening in a curtain. The water I've been drinking to avoid dehydration exacts its price, so I gotta go. I slip polar fleece pants and shirt over long underwear. I put on running shoes, not insulated boots, over wool socks I'm sleeping in. I put on a parka, but don't mess with its balky zipper. I pull on a stocking cap, gloves, sun glasses. Then I dash for the heated washroom, with flush toilets, about 150 feet across the ice. The –20°F (–29°C) cold numbs in my feet when I get there.

Polar Talk

Altitude sickness is a variety of symptoms resulting from the body not obtaining the oxygen it needs in lower-pressure air at high altitudes. Effects vary widely, and the condition can be deadly. The South Pole medical center has medications and a portable hyperbaric chamber, known as a Gamov bag, which a victim can be put into and the air pressure increased.

During recent summer seasons, the Pole was busy with several LC-130s arriving each day with materials to build the third U.S. station there. Snow had buried buildings erected in 1957 when the current station, which features a geodesic dome, opened in 1975. The dome shelters the buildings people live and work in from wind and snow. They look like heavy-duty, refrigerated trailers because they had to fit into an LC-130 to get to the pole. The doors are like those on walk-in freezers, except here the frozen food is stored outside in the cold air.

In the mid-1990s, the U.S. Antarctic Program External Panel looked over all of the U.S. Antarctic stations and said: "The U.S. would not send a ship to sea or a spacecraft to orbit in the condition of some of the facilities in Antarctica, particularly the one at the South Pole." Construction of the new station began in 2000 and is due to be done in 2007, at a cost of about $135 million.

Ice Chips

The 300 Club is a South Pole winter tradition. You can join only when the outside temperature drops to –100°F (–73°C). You sit in the sauna until its thermometer reads 200°F (93°C). Then, you put on sneakers or boots, maybe a face mask to keep frigid air out of your lungs, and wearing nothing else, run outside (some make it to the pole) and then back to the sauna—taking a couple minutes for the 300°F (166°C) drop and increase.

From November until late January, maybe 200 people will be at the pole any night, with most living in Jamesway huts at the "summer camp." In January, those who aren't going to spend the winter begin leaving. LC-130s pump fuel they won't need for the flight back into the Pole's tanks to power the generators that keep Pole residents alive. Once the last airplane leaves in February, the men and women at the pole won't be able to leave until October.

Travel Requires Attending Happy Campers School

U.S. Antarctic Program participants must complete a survival course, nicknamed "Happy Campers" school, before traveling away from McMurdo. You begin with a lecture on staying healthy in really cold weather and then ride out to the Ross Ice Shelf where you learn to build a snow shelter, pitch tents that fierce winds wont blow away, and how to use the stoves and other survival equipment carried on every vehicle that travels far from a station. To ensure you get a taste of the real thing, you spend the night in a snow shelter or tent you put up.

Explorations

Dehydration is a major polar hazard. In its dry air, you lose more water than you think. Everyone traveling in Antarctica is supposed to carry a water bottle. When you're out in the cold, you keep it inside your parka to keep the water from freezing. When you're camping, you fill the bottle with hot water and put it in your sleeping bag to help warm your feet. Thus, when you wake up you have a bottle of water, not ice.

Rescue Team Takes a Dangerous Trip

In December 1993, four Norwegians headed for the South Pole to retrieve the tent that Roald Amundsen had left there in 1911, to be displayed at the 1994 Lillehammer, Norway, Winter Olympics. As they headed south, two of their snowmobiles fell into crevasses about 575 miles from the pole. Two of the four were fine, one was injured, and the fourth disappeared into a crevasse.

Four members of the Rescue Team from McMurdo and New Zealand's Scott Base, led by Steve Dunbar, an American, landed in a Twin Otter airplane as close as possible to the Norwegians' camp. The rescuers, who were roped together, fell into crevasses 20 times during the 4-hour, 2-mile trip to the camp. Dunbar rappelled about 120 feet into a crevasse, to where it was about 8 inches wide, where he saw the arm of the dead victim about 5 feet below him, but he couldn't recover the body.

Crevasse Caution

Even experienced polar travelers have difficulty seeing crevasses because snow bridges can hide them. A bridge looks like solid snow but collapses when you step on it or drive a vehicle on it. Travelers connect themselves or their vehicles with ropes. The rope arrests a fall into a crevasse.

How to Find a Job in Antarctica

Raytheon Polar Services handles the day-to-day operation of U.S. Antarctic stations and research ships. Lori Boruch, the company's human resources director, says if you are a physician, a nurse, or a skilled plumber, electrician, sheet metal worker, or power plant mechanic, and can pass the physical exam, Antarctica wants you. Raytheon hires around 1,000 people to work during the October through February summer and another 250 or so on 8- to 12-month contracts that includes spending a winter at McMurdo, the South Pole, or the Palmer Station. The jobs pay from $300 to $1,400 a week plus transportation to Antarctica and room and board while there.

You Could Always Be a General Assistant

If you don't have one of the needed skills, the best bet is to apply for a job as a general assistant, or GA, which some say stands for "Good for Anything," which is a pretty good description of the job. "We've hired burned-out lawyers for general assistant jobs," Boruch says. As you can imagine, the competition is tough. "We get almost 6,000 unsolicited resumés a year. The cover letter should show you've done some research and you really want to go to Antarctica. If you say you want to go to Antarctica to see the polar bears, we toss the resumé away."

For Americans, other possibilities are joining the U.S. Coast Guard to sail on one of the ice breakers that go to Antarctica, and also the Arctic Ocean, or joining the New York Air National Guard's 109th Airlift Wing based at the Schenectady Airport, which flies LC-130s in Antarctica and Greenland.

Ice Chips

To work for Raytheon Polar Services, you need to be a U.S. citizen with a passport, or a resident alien. Information about jobs with other national Antarctic programs is available on the website for the Council of Managers of Antarctic Programs in the Bibliography.

Students and Teachers Head for the Poles

Adventurous science teachers who would like to work in a polar region should look into the Teachers Experiencing Antarctica or the Arctic (TEA) program, run by the National Science Foundation, the American Museum of Natural History, the Army's Cold Regions Research and Engineering Laboratory, and Rice University.

If you are a student who wants to be a polar scientist, this book's information about research should give you a good idea what to study. Joe Grzymski, a young scientist, sums it up: "Antarctica is an incredible place to work and to learn. It requires some major sacrifices and precautions, but it is rewarding beyond imagination."

The Least You Need to Know

- Despite technology, Antarctica remains dangerous.
- The United States is building a new South Pole Station to replace the outdated 1970s station.
- Survival school is required for Antarctic travel.
- Physicians and people with trades such as plumbing are always needed to work in Antarctica.
- Becoming a general assistant is a good way for someone without special skills to work in Antarctica.
- Antarctica has opportunities for teachers and scientists.

Chapter 19

Travel to the Polar Regions

In This Chapter

- Almost anyone can visit the polar regions today
- You don't need sled dogs to reach the North Pole
- Getting to Antarctica isn't easy, nor inexpensive
- Most Antarctic tourists visit the Peninsula
- You need to carefully plan a polar trip

Until the last half of the twentieth century almost everyone who traveled to either of the earth's polar regions was either an explorer or had a polar job, such as at a weather station. Today, polar tourism is a large and growing business. You can find a polar trip that matches the magnitude of your spirit of adventure and your checking account. In this chapter, we will give you some ideas of the kinds of polar travel that are available, and a few ideas for making sure you select the right trip. Before taking off on a trip, however, you should expect to do a good amount of research or put yourself in the hands of a trusted travel agent with experience booking the kind of trip you're thinking of.

Polar Regions Are No Longer for Explorers Only

A visit to the polar regions today can be as simple as having your picture taken next to the "Arctic Circle" sign on the highway north of Fairbanks, Alaska, or as elaborate as a $25,000 adventure trip that takes you to the South Pole for three hours.

Those who want to rough it, living in tents or youth hostels, can find many Arctic trips of their dreams, although getting there is likely to be expensive. Antarctica is out of the question for casual campers, because there are no "common carrier" flights or sailings. All the airplanes and ships going to Antarctica are either for those working for one of the national research programs or those on cruises or adventure tours. The only way of getting to Antarctica on your own is to sail a yacht, which some people do.

No matter where you go in the Arctic, expect high prices. Hauling fuel, food, sheets, and towels for a hotel or trinkets for gift shops is expensive. Hauling things to Antarctica is more expensive, but travelers aren't likely to notice. If you're on a cruise or adventure trip, the cost of room and board and most of your activities are included. (This is something you should verify before signing up for a trip.) Because Antarctica has never had a native population, you won't find any lovely but pricey folk art to buy. Small science station stores selling souvenirs are Antarctica's only shops. Just how much do you want to spend for a T-shirt, baseball cap, and postcards at the South Pole?

You Can Taste Alaska's Arctic on a Day Trip

For most Americans, the easiest and least costly way to visit the Arctic is to go to Alaska. Just don't expect to see igloos and people in furs when landing in Anchorage, which is a medium-size American city in a spectacular setting. Only the northern third of Alaska is in the Arctic.

You can drive to within a few miles of the Arctic Ocean from the "Lower 48" states, although this is a driving adventure that will take more time and planning than a trip, say, from Philadelphia to San Diego. First, you have to drive across western Canada on the Alaska Highway to Fairbanks. From here, you drive north on the Dalton Highway—which is not paved—to Deadhorse. The Dalton Highway is a narrow, gravel road with a lot of truck traffic and few places to buy gasoline.

> CAUTION
>
> **Crevasse Caution**
>
> If your young child happens to see "North Pole" on a map of Alaska, don't let him get too excited. North Pole, Alaska, is a suburb of Fairbanks and is a good 1,700 miles (2,735 kilometers) south of the real North Pole. The main attraction in North Pole, Alaska, is a Christmas-theme gift shop. And yes, Santa Claus does hang out there.

If you want to have your photo taken at the "Arctic Circle" sign, drive or take a tour about 190 miles north from Fairbanks to the Circle. If you're going to do this, think about going late at night near the summer solstice on June 20 or 21. There's really not much to see where the Arctic Circle crosses the Dalton Highway, but being there to see the midnight sun would make it a little more memorable.

To really go to the Arctic, hop an Alaska Airlines flight to Barrow, Alaska, on the Arctic Coast and see the true Arctic—tundra, flocks of birds you won't see at home, midnight sun every day from May 10 through August 2, and the first-rate Inupiat Cultural Center, where you can learn a lot about the Eskimos in this part of the Arctic. Even in the middle of summer you might see ice floating in the Arctic Ocean. You can arrange a trip out to the northernmost point in the United States at the end of Point Barrow—past the sign that tells you to watch out for polar bears. The big advantage of Barrow is that you can experience the Arctic without the expedition skills or guide you need elsewhere in Alaska's Arctic.

That's not the case elsewhere in Alaska's Arctic. Take the Gates of the Arctic National Park, for instance. The park's eastern edge is within five miles of the Dalton Highway, and its visitors' center at Coldfoot on the highway is open in the summer. But no road goes into the park; you'd have to hike. As the National Park Service says on its website: The park "is a remote wilderness and travelers should be fully competent in outdoor survival skills."

Ice Chips

The car rental agreement with the UIC Development Co. in Barrow says in part: "Do not haul any whale meat, waterfowl, fish, or game that has not been properly packaged and sealed. You will be charged a minimum cleaning fee of $50, and may be charged for actual damage resulting from carrying meat or fish. In extreme cases this may involve replacing the vehicle."

The Arctic Is Easiest to Reach in Europe

As far as getting to the Arctic goes, no place could be more different from Alaska's wilderness areas than the European Arctic, which includes northern Norway, Sweden, Finland, and European Russia. You can drive to the Arctic Ocean in Norway, although it's a long trip—1,370 miles (2,205 kilometers) from Oslo. For an Arctic travel bargain, take the train using your Norway Rail Pass, to Fauske and Bodo on the line paralleling Norway's Coast or Narvik on a line from Sweden.

You could travel north to Norway's Svalbard Islands, where several important 1920s and 1930s Arctic flights began and ended. After looking over the Svalbard Tourist Office website, listed in the bibliography, and reading what guide books have to say about strict environmental regulations and dangers of polar bears, you might want to opt for a cruise or tour to get you there and back.

Crevasse Caution

The Svalbard Tourist Office warns that because "polar bears are essentially everywhere" you should "bring a weapon with you on all trips outside of the settlements. Large-bore rifles are absolutely the best defense against polar bears. Weapons may be rented locally. If you are going to spend the night in a tent or under the open sky, the campsite should be secured with tripwire warning flares."

For Remoteness, Go to Canada's Eastern Arctic

Getting to Canada's Ellesmere Island National Park in the Territory of Nunavut takes some doing, but for the dedicated wilderness traveler, the journey is worth it "Up here (Ellesmere Island National Park), above the treeless expanse where the tallest vegetation barely reaches your knees, the red and green Northern Lights dance with such vigor you can hear them crackle with energy," said Margo Pfeiff, a veteran Arctic trekker. "From 2,000 meters (6,562 feet) up, icebergs dot the inky ocean like grains of rice and the tundra is polka-dotted with a kaleidoscope of pothole lakes from turquoise to indigo." If you go here, you need a guide and you need to be comfortable in the wilderness. It is not a "soft" wilderness adventure.

Nunavut is one of the world's last great wilderness areas, and it teams with wildlife ranging from arctic char, narwhal, beluga whales, walruses, and seals, to herds of musk oxen, caribou, and even Arctic hare. Packs of wolves and solitary Arctic foxes lead nomadic lives on Nunavut's tundra, while at least 30 species of birds are seen in the skies over Ellesmere Island. Nunavut is also the scene of much of the sad history of the search for the Northwest Passage (see Chapter 11). You can visit several historical sites, especially near Gjoa Haven, where Roald Amundsen paused for almost two years during his successful 1903–1905 voyage through the Northwest Passage. The Franklin expedition of 1845 was lost near what is now Resolute, which is also an air hub for this part of Canada.

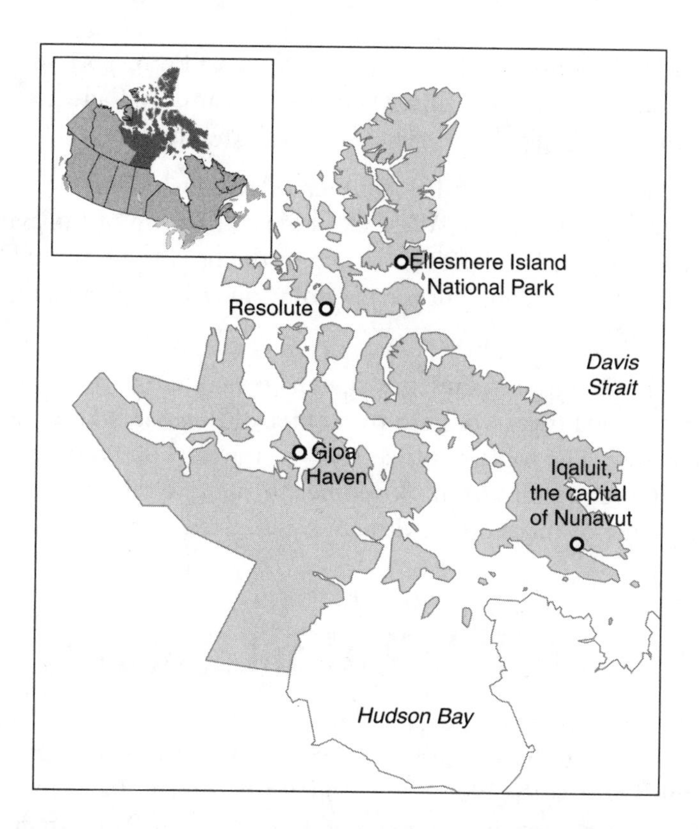

Nunavut Territory.

Greenland, Hard to Reach but Worth It

Air travelers between Europe and North America sometimes catch a glimpse of Greenland's Ice Sheet, and that's the nearest most people ever come to the world's biggest island. Unfortunately for North Americans, Greenland's tourist efforts are focused on Europe, where regular air service is available from Copenhagen, Denmark, to Kangerlussuaq, the island's air hub. An American who wants to go to Greenland should expect to pay much more than he'd pay for a trip to Europe to get to an island that's halfway across the Atlantic. The Greenland travel information website listed in Appendix C has a link to a list of U.S. travel agents who handle Greenland travel. For most Americans, the best way to see some of Greenland is to take one of the growing number of cruises that visit ports along Greenland's coast.

Although Greenland has tundra, fjords, impressive mountains, and Arctic wildlife such as musk oxen, the Ice Sheet makes it truly special. Tourist offices in several communities can arrange trips over the Ice Sheet by helicopter or hiking or even dogsled trips on the ice. For those who like to sip a beer, dine on grilled musk ox steak, and watch the ice sheet melt, the Kangerlussuaq Tourist Office will transport you to the edge of the Ice Sheet for a picnic. Sometimes visitors will hear creaks and groans from the ice become a rumble as a huge chunk of ice falls onto the tundra.

Anyone who can should see Greenland's Jacobshaven Glacier, which flows down the Ice Fjord into Disko Bay on Greenland's West Coast near the town of Ilulissat. One of the unique experiences Greenland offers is sitting in the bar of the Arctic Hotel on the edge of town, say at 1 A.M., looking out the picture windows as icebergs, illuminated by the midnight sun, float by.

The hotel and bar would be right at home in Copenhagen—after all, Greenland is Danish—but a short walk will bring you into the center of Ilulissat, where the Inuit heritage is very much alive. The tourism office says the town's 4,000 residents have at least 2,500 sled dogs. If you're there from November through April, the local tourist office can arrange for some of those dogs, with an experienced driver, to take you for a ride.

How to Reach the North Pole Without Sled Dogs

You no longer have to spend years in the Arctic, maybe losing toes to frostbite and learning to drive a dog team, to have a good shot at making it to the North Pole. The main qualifications for reaching the North Pole today are a spirit of adventure, a tolerance for a little—but not too much—danger, and lots of money. You can cruise to the North Pole on a Russian icebreaker, or you can fly there from Canada or Russia.

An Icebreaker Will Take You to the Pole

The collapse of the Soviet Union opened up both Arctic and Antarctic tourism by making Russian *icebreakers*, including nuclear-powered icebreakers, available for lease to western tour companies. An icebreaker is needed to reach the North Pole, and some parts of Antarctica, such as McMurdo Sound. Such cruises are pricey and the western companies that run them strive to meet the standards of comfort high-end travelers expect.

Polar Talk

A ship designed to break through pack ice is called an **icebreaker**. Such ships have extra-strong hulls and more powerful engines than ordinary ships of the same size. They break the ice by riding the front of the ship up onto the ice, letting the weight of the ship break it.

The cost of a two-week cruise to the North Pole, leaving from Oslo, Norway, starts at around $16,000 a person. A variety of other Arctic cruises are available at a wide range of prices. At the low end, you can take one of Norway's coastal ships from Bergen in southwestern Norway along the coast, in and out of fjords, to Kirkenes, on the Arctic Ocean near the border with Russia. This 7-day trip starts at around $2,000 per person. For around $10,000, you can cruse on an *ice-strengthened ship* along Amundsen's Northwest Passage route.

Polar Talk

An **ice-strengthened ship** is a ship with a strong hull designed to move through pack ice that isn't solid enough to require an icebreaker. The ship can withstand collisions with ice floes, but not break solid ice.

Fly to the North Pole and Jump Out

You can fly to the North Pole on trips of various lengths, and with a wide range of activities such as sky diving, running a marathon, or skiing the last degree of latitude, which is 69 miles (111 kilometers), to the pole. Trips leave from Russia and Svalbard, Norway, using Russian airplanes and helicopters and land at Russian camps on the ice, where you might spend a couple hours or a few days, depending on the tour. Trips from Canada leave from Resolute, Nunavut, with a stop at the Environment Canada weather station at Eureka to wait for the right weather for a dash to the North Pole.

Getting to Antarctica Isn't Easy, Nor Inexpensive

In the January 14, 2001, issue of the *Antarctic Sun*, published at the U.S. McMurdo Station, Josh Landis, a *Sun* writer, noted: "Any traveler that makes it this far south probably has a strong affinity for Antarctica and a large disposable income." His story was about a visit of the cruise ship *Kapitan Khlebnikov*, one of the icebreakers now in the polar cruise business. The tourists who spent a couple hours touring the McMurdo station had paid from $9,950 to $19,750 per person for the three-week cruise, not including the airfare to Hobart, Australia, to board the ship.

For that price, however, they were seeing things that few people ever see, such as the Ross Island huts built by the Heroic Age explorers Robert Scott and Ernest Shackleton and the Royal Society Range of the Transantarctic Mountains across McMurdo Sound on one side of the ship and Mount Erebus on the other side.

"It's really fun, just incredible. I can't even think I'm here," Maria Eriksson, age 11, of Helsinki, Finland, told the *Antarctic Sun*, surely echoing what most of the adults were thinking. During a stop at Terra Nova Bay to the north on the Ross Sea, Maria sat down near emperor penguin chicks and one walked to within a foot of her. "Its down was like rabbit fur," she said.

Most Tourists Visit the Antarctic Peninsula

You can cruise to Antarctica for maybe a third of what those aboard the *Kapitan Khlebnikov* paid, but you won't get as far south. In fact, the great majority of those who cruise to Antarctica visit the Peninsula and Southern Ocean Islands. The International Association of Antarctica Tour Operators estimates that during the 2002–2003 season about 16,000 tourists were on 252 cruises to the Peninsula while only 332 passengers were on the five cruises that visited the Ross Island area or sailed around the continent.

Cruises that visit Ross Island or circle Antarctica usually leave and return to Australia or New Zealand. Peninsula cruses usually depart from southern Argentina or Chile.

Until recent years, most cruises to Antarctica have been on relatively small, ice-strengthened ships carrying fewer than 150 passengers. The ships stop maybe two or three times a day for passengers to go ashore, closely guided by the ship's expedition staff, to explore and maybe see a penguin colony. Some ships offer activities such as scuba diving. Today, however, a few large ships carrying more than 1,000 passengers cruise near the peninsula but have not included shore excursions.

If you are thinking of taking an Antarctic cruise, you need to decide first how much you want to spend and whether you want to gaze at Antarctica's amazing landscape of rocks and ice from a distance, or do you want to walk around on it, get close to some penguins—being sure not to annoy them. Another point to consider: Will the ship take you south of the Antarctic Circle? Although the Circle is merely a line on a map, you might not feel as through you've really been to Antarctica, no matter how much ice and how many penguins you see, unless you cross the Circle.

You Can Go All the Way for $25,000

For the ultimate in "Been there, done that" T-shirts, you can pay Adventure Network International $25,000 or more per person, to go to the South Pole, where the store at the Amundsen-Scott South Pole Station will sell you a T-shirt, and put a "South Pole" postmark on your post cards. The price includes a flight from Punta Arenas, Chile, to the company's camp at Patriot Hills about half way from the Antarctic Coast to the South Pole, and about 10 days at Patriot Hills, skiing, hiking, and taking snowmobile trips in the area.

The highlight, of course, is a flight to spend about three hours at the South Pole, learning about the research done there, buying T-shirts and other items, and having your "hero" photo taken at the South Pole marker. Like some recent visitors, you might want to play the saxophone by the pole marker or dance with your wife on the South Pole snow.

Flights Give You Antarctica Without a Parka

If you really, really hate the cold, but would love to see Antarctica, a travel company in Australia offers a way to do it: Fly from Australia in a Qantas Airways Boeing 747 over the continent without landing. During the 2002–2003 season, Croydon Travel in Croydon, Australia, offered three flights, including one on New Year's Eve, which included a jazz band and Champagne.

The 12-hour flights depart from and return to Australian cities such as Sydney and Melbourne and include about 4 hours over Antarctica. They don't get as far south as the South Pole, but the pilots can select cloud-free areas to fly over. Lectures and videos help you pass the time to and from the continent.

Polar Travel Isn't a Trip to Disneyland

As you prepare to take a polar trip, keep in mind that you are about to set off to a part of the world where extreme weather is common; to a place that probably does not have the infrastructure and comforts you are used to at home. With few exceptions, you are not going to find highways, much less train service. If you go to the polar regions, you are almost surely going to travel by airplane or ship. You should be prepared for flight delays; take a good book, and consider delays part of the adventure of a lifetime. Books listed in Appendix C will give you ideas for reading material to take on your trip.

Even if you are taking a cruise and don't plan to man-haul a sled across the ice or climb any mountains, you still need to think about your health. Many polar cruises require you to supply medical information. Being anything but completely honest is not likely to hurt anyone but you.

Cruise ships have physicians and basic medical facilities, but you will be far away from a major medical center. If you have any doubts, discuss your trip and your doubts with your physician before booking a trip. No matter how healthy you are, if you are taking a cruise to Antarctica, you are more likely to suffer from seasickness than on any other ocean—although some cruises are lucky and find the Drake Passage calm. Still, you should have your favorite seasickness remedy when you board.

For a Good Time, Buy the Right Trip

How well your polar trip turns out depends on how well you've planned for it. Planning begins with learning all you can from books and the Internet. (You've probably learned by now not to believe everything on the Internet.) Even after you've done your research, you should consider finding the right travel agent, advises Talula Guntner, a professor of travel and tourism at Northern Virginia Community College in Annandale, Virginia. "You want to build a relationship with a travel agent like you build with a hairdresser. The hairdresser knows how you like your hair cut," she says. Your travel agent should be just as good at knowing what kinds of trip will make you happy.

How do you find the right travel agent? Guntner says it's like finding the right physician. Ask your friends, meet the agent, and see if you think he or she knows the business and is interested in finding the right trip for you. Make sure the agent is a member of the American Society of Travel Agents (ASTA). Among other things, members must carry "errors and omissions insurance." Guntner says: "The main thing is to find someone you trust, who's ethical, and does the right thing."

CAUTION

Crevasse Caution

If you're booking an expensive polar tour, you want to make sure your money doesn't disappear before you enjoy the trip. One way to do this is to book with a company that's a member of the U.S. Tour Operators Association (USTOA). Members are bonded, which means you have a better chance of getting your money back if the company becomes unable to provide the tour you paid for. Even with this protection, however, you should buy insurance against an operator going belly up, because the USTOA bond might not be enough.

When you buy a cruise or tour, you have to make a deposit. Before doing this you should carefully look into travel insurance, and this can get complicated. You can buy trip cancellation insurance and medical insurance. Guntner says cancellation insurance or a cancellation protection plan offered by cruise lines often will let you cancel for any reason up to the time to leave, but may or may not cover you on the trip. But, she says, a tour company or cruise line can't insure against its own bankruptcy. For that you need a policy sold by someone else.

You should also look into medical emergency insurance, especially a policy that would bring you home if you suffer a serious injury or illness. This is part of some trip cancellation and evacuation policies.

Helping you select the right trip insurance could be one of the most important services a travel agent could offer. Even with the best travel agent helping, you should read and understand exactly what the policy does and does not cover. You should make sure that anything anyone tells you is covered is written into the policy. This is a time to ask questions: What reasons for cancellation will the policy cover? What happens if I become ill on the trip? If you read the fine print only once in your life, it should be the fine print in a travel insurance policy for a big trip.

The Least You Need to Know

- You don't have to be an explorer to visit the Arctic or Antarctic.
- Alaska offers a relatively easy way to see the true Arctic.
- Travel in the European Arctic is not much different from travel elsewhere in Europe.
- Greenland is expensive to reach, but can be worth it.
- You can travel to the North Pole by luxury icebreaker or airplane.
- About 85 percent of Antarctic tourists visit the Peninsula; visiting other areas is very expensive.

Part 5

Today's Polar Science

The science that's going on in the polar regions today takes us from the tops of ice sheets to the bottoms of cold oceans and even to a lake hidden under 2½ miles (4 kilometers) of Antarctic ice. We'll see what scientists are trying to learn about the many creatures that live in the Arctic Ocean and how other researchers are tracking ocean currents and temperatures deep under the ice at the North Pole. We'll also look at what researchers are learning about the ozone hole, penguins, rocks, and fossils.

STUDYING ASTRONOMY IN THE ANTARCTIC IS LIKE MY TIME IN SCHOOL. A DAY SEEMS TO LAST THREE MONTHS AND ALL I'D DO IS STARE OFF INTO SPACE.

Chapter 20

Antarctic Skies and Stars

In This Chapter

- Antarctica is a perfect place for sky studies
- How the sun causes auroras
- A visit to the frontiers of astronomy helps us understand why astronomers go to Antarctica
- Instruments at the South Pole look back to when time began with a big bang
- A big, new South Pole telescope could answer some big questions

The high Polar Plateau around the South Pole is more than 9,000 feet (2,700 meters) above sea level, and its cold, thin air makes it a sky watcher's dream. In this chapter, we will first marvel at the curtains of auroras dancing in the dark. Then we'll turn our eyes to the clear, unblinking light from the stars, which Polar Plateau sky watchers and telescopes see with a clarity found nowhere else on Earth. In addition, the Plateau is the best place on Earth to study all the forms of energy from the far reaches of space that our eyes don't detect. This allows astronomers' instruments on the Plateau to view *galaxies*, *black holes*, exploding *supernovae*, and the faint glimmer of time's beginning.

Polar Talk

Galaxies are giant assemblies of stars, gas, and dust. Most of the visible matter in the universe is found in galaxies. A **black hole** is an object so collapsed in on itself that not even light can escape from its intense gravitational field. A **supernova** is an exploding star that can become billions of times as bright as the sun before gradually fading from view.

Polar Nights Glow with Nature's Neon Lights

The eerie glow of an aurora bathes a ring around the Antarctic and also the Arctic when the sun turns stormy. The display dips down to the surface in a ring around Earth's magnetic poles and we earthlings who are close enough watch the show. The energy fueling auroras originates with the sun, transported in its wind. South Pole watchers gaze at "southern lights," or *aurora australis*. At the opposite end of the earth, watchers see the same kind of phenomenon but call it the *aurora borealis*, or the northern lights.

Most of the humans who have ever seen an aurora have seen the northern lights because people live where the northern lights glow. The aurora is seen most often below the Northern and Southern hemisphere *auroral ovals*. These are centered on the North and South geomagnetic poles and are oval, with the side away from the sun pushed a little farther toward the equator and often a little wider. When the sun is calm, the auroral ovals are thin and auroras are seen only in or near the polar regions. At times, however, the sun kicks up its heels and we see auroras farther from the poles.

Polar Talk

The luminous phenomenon in the Southern Hemisphere created by energetic particles from the sun hitting atoms of atmospheric gasses is called the **aurora australis.** The luminous phenomenon in the Northern Hemisphere created by energetic particles from the sun hitting atoms of atmospheric gasses is called the **aurora borealis.** An **auroral oval** is an oval band around a geomagnetic pole where auroras are seen most often.

Unless you want to get a job that keeps you in Antarctica over the winter (see Chapter 18), the best way for most people to get good views of the aurora is to travel to interior Alaska or to northern Canada, where the northern lights can be seen close to 200 days a year. In Europe, the northernmost part of Scandinavia offers the best auroral viewing, but in the winter clouds can often block the view. Probably the best time to go is in March or September when the nights are 12 hours long, but you miss the coldest weather.

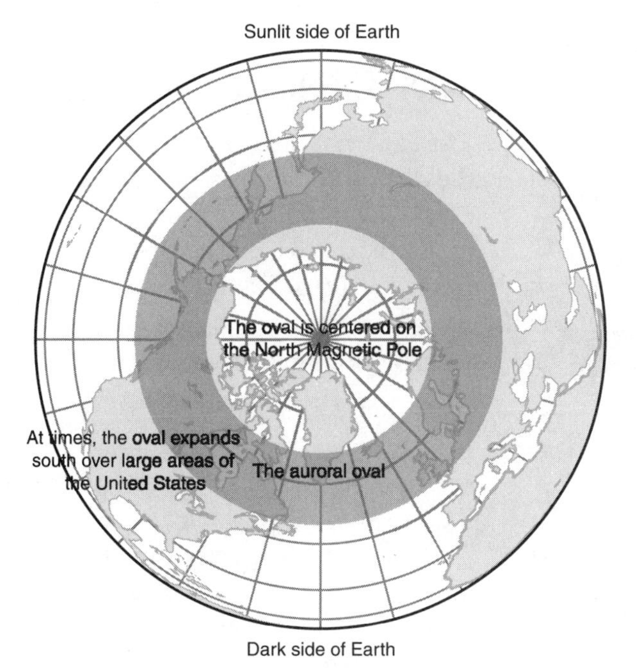

A typical Northern Hemisphere auroral oval.

A Privileged Few See the Southern Lights

Antarctica—home of the Southern Lights—is home to few humans and most of those leave at the end of the summer as the sun is going down. Occasionally a curtain of light will dip north far enough for Australians to gasp and murmur their delight.

Here's how Simon Hart, an astrophysicist who spent the Antarctic winter of 1996 at the South Pole, describes the aurora in his web diary:

> For a few days each month, we were treated to fantastic displays lasting several hours. Initially, a long ribbon appears over the dark sector, snakes its way to us, passes right overhead, and writhes across the sky north to the opposite horizon. The approaching ribbon turns into a rippling curtain of ghostly green light, sometimes tinged neon-pink. Directly overhead, it looks like hundreds of light fingers waggling furiously—a silent waterfall, constantly raining down but never reaching the ground. At times, it seemed close enough to touch but always just out of reach. The most intense displays cast shadows on the snow.

Polar Talk

The **dark sector,** which is half a mile from where people live, is the location of many of the South Pole's scientific instruments, and is so named because there's no artificial radiation (visible or infrared) to contaminate observations.

Polar Talk

An **electron** is a subatomic particle having a negative electric charge. A **proton** is a positively charged subatomic particle having a mass 1,836 times that of the electron.

Polar Talk

The **magnetosphere** is a region surrounding Earth, extending from about 60 miles (100 kilometers) to several thousand miles above the surface, in which charged particles are trapped and their behavior is dominated by Earth's magnetic field.

Auroras Begin with the Sun

The sun's extremely rarefied outer atmosphere, called the corona, produces a solar wind by continuously expanding into interplanetary space. Unlike the wind on Earth, this wind is not moving air. Instead the corona's expansion creates a steady stream of mostly *electrons* and *protons* flowing outward like a child's picture of sunrays.

As the solar wind sweeps past a planet with a magnetic field, such as Earth, it distorts the field and creates a protective shield that surrounds the planet. This region, called a *magnetosphere* and shaped like a comet's tail trailing away from the sun, traps charged particles. The solar wind interacts with the magnetosphere and ultimately with Earth continuously. But that's not what causes auroras. The solar wind just forms the magnetosphere.

Our Atmosphere Glows Like a Neon Sign

As the particles descend, following the earth's magnetic lines of force, they occasionally collide with atoms in the earth's upper atmosphere and some of the air atoms glow as a result. An atom glows because the charged particle that hit it transfers kinetic energy to the atom. The atom often dumps the extra energy by emitting light and returns to its normal energy state. It glows like a neon atom inside a neon sign.

Different atoms glow different colors to form vibrant, many-hued rings, some continent-size or larger over Earth. Bombarded ionized molecules of nitrogen shine blue. Oxygen atoms hit by incoming charged particles 200 miles high glow red, the rarest aurora color. At about 60 miles, glowing oxygen produces the most common color: a brilliant yellow-green. The different colors depend on how often particles hit an oxygen atom; at lower altitudes where there are more atoms, they are hit more often, creating the green glow.

Sunspots Flare to Cause Auroras

Regions of strong vertical magnetic fields on the sun (called solar active regions or groups of *sunspots*) are probably the ultimate cause of the big auroras. Sunspots, which are places where the sun's surface is cooler than the surrounding area, appear to be regions where magnetic loops from inside the sun pierce the surface. Activity connected with sunspots creates *solar flares*, which are a sudden intense brightening of a small part of the sun's outer surface. It's a concentrated, explosive release of energy. A flare develops in a few minutes and may last hours. It may flash on and off a few times and it emits a gust of energetic, charged particles into interplanetary space.

Polar Talk

Solar flares are sudden eruptions of hydrogen gas on the surface of the sun, usually associated with sunspots and accompanied by a burst of ultraviolet radiation that is often followed by a magnetic disturbance. **Sunspots** are any of the relatively cool dark spots appearing periodically in groups on the surface of the sun that are associated with strong magnetic fields.

An unusually large solar gust hit Earth's magnetosphere in 1909 and created an aurora seen in Singapore in Southeast Asia, on the earth's geomagnetic equator.

Explorations

In addition to the familiar geographic poles at the ends of the globe's spin axis and the magnetic poles at which compasses point, Earth also has North and South geomagnetic poles. The geomagnetic poles are more theoretical than the others, being based on the current International Geomagnetic Reference—a model of the earth's main magnetic field. The model views Earth's magnetic field essentially as like that of a bar magnet. The geomagnetic poles are the ends of the magnet. They are symmetric. Earth's field changes constantly; so geologists revise the model every five years and track how the field changes.

Antarctica's Frigid Sky Makes Astronomers Glow

Antarctica delights astronomers for several reasons. To begin with, the colder the air, the less water vapor it has. Thus starlight over Antarctica's Polar Plateau never appears hazy. Also Antarctica's air stays still more often that in warmer places. In the winter, the air right next to the ice and snow is extremely cold, but if you sent up a weather balloon, its thermometer would record warmer temperatures for a while as it ascended. This is known as an inversion. Inversions calm the atmosphere by suppressing the up and down air movements known as convection, which cause the stars to twinkle. Most people find this twinkling charming, but not astronomers when they're at work.

Antarctica's clear, calm sky allows stars to shine bright steady light into a profusion of telescopes—optical, radio, *infrared*, and *submillimeter*. Orbiting satellites see better than any on Earth, but at astronomical costs. Building telescopes in Antarctica is costly, but not nearly as costly as sending them into space.

Finally, if you are at the South Pole the stars never set, they circle you. Astronomers can keep their telescopes and other instruments trained on the same spot in the sky 24 hours a day.

Polar Talk

Infrared (heat) astronomy is the study of space radiation with wavelengths beyond the red end of the visible spectrum, from 800 to 1 millimeter approximately. **Submillimeter astronomy** is a branch of astronomy that covers radiation wavelengths from 0.3 to 1 millimeter approximately. It is important in the study of giant molecular clouds, which are cool, dense regions of interstellar matter where atoms tend to combine into molecules.

Pole Instruments Track Far-Away Space Violence

To understand why astronomers spend so much money building special telescopes in Antarctica and work so hard in such a cold place, we'll take some excursions away into the frontiers of today's astronomy in the rest of this chapter. Each of these side trips, however, will bring us back to the South Pole.

In the universe that polar astronomers are trying to understand, black holes collide, stars rip apart, *dark energy* and matter lurk. When something enormously violent happens in space, the explosion releases radiation and particles of all kinds: light, radio waves, *cosmic rays*, and *neutrinos*. Instruments at the South Pole—telescopes staring into space and arrays frozen deep within the polar ice—receive the messages.

Polar Talk

Dark energy is unknown energy that may make up 65 percent of the universe. The energy is repulsive and shoves galaxies away from each other at an ever-increasing speed. Dark energy only shows up over significant fractions of the observable universe. **Cosmic rays** are high-energy particles that move through space at close to the light speed and bombard Earth from all directions. **Neutrinos** are elementary particles with zero charge and spin. They have very small mass and weak interactions with matter.

AMANDA and IceCube Detect Cosmic Crashes

A neutrino telescope at the South Pole is ready to capture the next violent event in the Milky Way—and soon, in far-away galaxies. The telescope, called AMANDA (the Antarctic *Muon* and Neutrino Detector Array) doesn't stare up into space. Instead it looks down, north—through our entire planet—and then out into northern space. AMANDA doesn't see light; it detects barely discernable neutrinos. It uses the earth to filter out the effects of cosmic rays so its detectors trigger only on neutrinos.

Polar Talk

A **muon** is an elementary particle with the same charge and spin as an electron but with a mass 207 times bigger.

AMANDA sits deep in Antarctica's clear ice—1 mile (1.5 kilometers) below the surface. Neutrinos interact (rarely) with AMANDA's ice. Occasionally, though, an invisible neutrino careens through the ice and crashes into a proton or an electron. A heavy electron (called a muon) emerges from the wreckage and zooms off perhaps hundreds of yards. The muon gives off a shock wave of blue light as it zips through the ice. AMANDA detects this faint light—the debris of a neutrino hit.

Matthew Ribordy and Yulia Minaeva from the University of Stockholm review data from the Antarctic Muon and Neutrino Detector Array (AMANDA) at the South Pole Station.

(Courtesy National Science Foundation, Melanie Conner)

Any device, such as AMANDA, must watch a huge hunk of ice to snag any neutrinos at all. That's why AMANDA is large: three times as tall as the Eiffel Tower, a cylinder 3,300 feet (1,000 meters) high with a 660-foot (200-meter) diameter base. Because the telescope sees only a tiny blue light whenever a neutrino deigns to interact with the ice—the telescope must be dark and the ice clear. That's why AMANDA is buried in the dark far below the surface and uses the clear ice of Antarctica.

As big as AMANDA is, it isn't big enough to detect neutrinos from other galaxies. AMANDA has detected high-energy neutrinos produced in Earth's atmosphere near the North Pole. Astronomers are planning a bigger array, called IceCube, to replace AMANDA in 2008. IceCube would house about a quarter of a cubic mile (1 cubic kilometer) of ice—almost 200 times the volume of AMANDA and will peer deep into the cosmos. It would study some of the most violent and least understood objects in the universe.

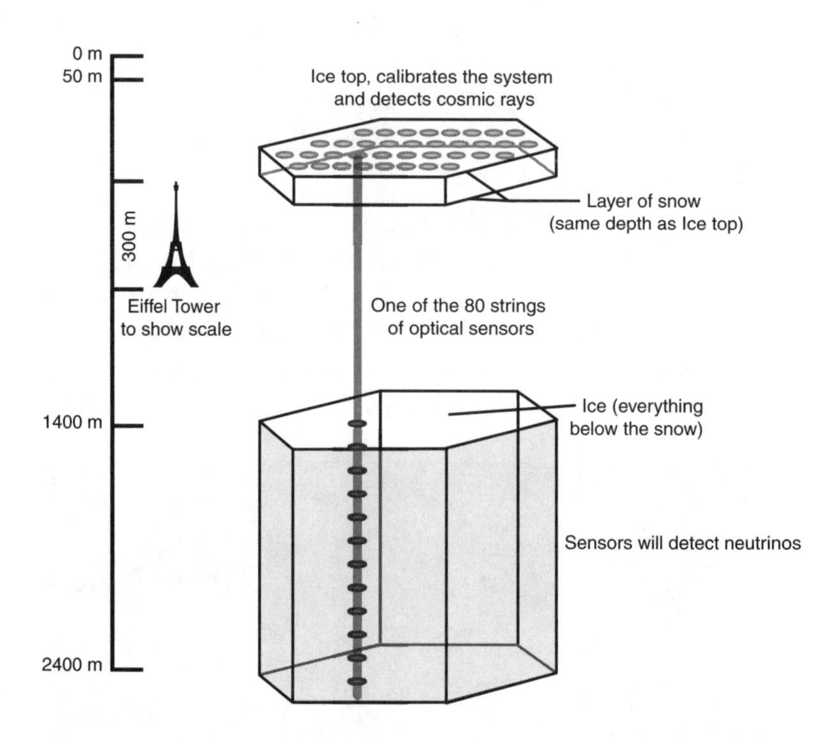

A look into the ice at IceCube.

Neutrinos are neat messengers from afar. They come straight to us—interstellar magnetic fields don't deflect them and intervening matter doesn't absorb them. So they can make it here from outermost reaches of the universe. What's more, AMANDA and IceCube sense them in such a way that it can track where the neutrinos came from. That's one huge "Now I understand!" Knowing where a cataclysmic event occurred eons ago, we can train Earth's greatest telescopes upon it, study it, and uncover its secrets.

South Pole Telescopes Get the Data

On September 19, 2002, University of Chicago astrophysicists announced a momentous discovery. Using a radio telescope, called the Degree Angular Scale Interferometer

(DASI), at the South Pole Station—they measured a tiny *polarization* of the fossil radiation that originated from almost the beginning of time, 14 billion years ago. They found that the *microwave* background radiation in adjacent patches of sky vibrates in slightly different directions. That may sound less momentous than it is. It means, though, that the Big Bang theory checks. The model predicts just such vibrations and … bingo! We detected them.

Polar Talk

Microwave radiation is a high-frequency electromagnetic wave, one millimeter to one meter in wavelength, between infrared (heat) and short wave radio wavelengths. **Polarization** is a state in which light-wave vibrations line up in the same direction.

The Martin A. Pomerantz Observatory (MAPO) and (to the right) the DASI telescope at the South Pole.

(Courtesy National Science Foundation, Melanie Conner)

Fossil Radiation Cooled and Stretched

The light that started toward us 14 billion years ago at the birth of the universe was the same color as the cooler giant stars are now. But today, we can't see light visible so long ago because expanding space changed the light as it zapped here. The ballooning universe stretched the light a thousand fold from visible light to longer-wavelength radiation: microwaves—similar to what we cook with in microwave ovens. The ancient photons still exist in vast quantities. In fact they make up about 1 percent of the static noise picked up by a TV antenna.

That's the radiation that the radio telescope, DASI, received at the South Pole in 2002. By examining its polarization, scientists can determine not just how lumpy the early universe was, but how its particles moved. That's a powerful handle on how the universe formed and how it's evolving.

The Next Big Leap Forward Is a Huge Telescope

In 2000, a survey of astronomers reported that a new South Pole telescope, similar to DASI but better, was one of the top 10 research priorities. The National Science Foundation hopes to install a $16 million telescope in 2009. The mammoth instrument, 26 feet (8 meters), in diameter, would dwarf other telescopes and many structures at the Amundsen-Scott South Pole Station. If it's built, astrophysicists hope to determine if dark energy exists.

The new instrument would look at a shorter wavelength of the cosmic radiation and be able to detect finer variations and weaker signals. With it, astronomers would chart clusters at different distances and ages from us, giving us clues about how dark energy influences the growth of galaxy clusters. That information would probably tell astronomers how the universe evolved.

The Least You Need to Know

- Antarctica's cold, clear air makes for great sky watching.
- Streams of particles interacting with Earth's magnetic field cause the auroras.
- The northern lights are seen over the populated Northern Hemisphere, but you have to be in Antarctica to see the southern lights most of the time.
- A device buried in the ice under the South Pole detects tiny particles known as muons and neutrinos, and helps scientists understand the universe.
- In 2002, astrophysicists announced that a radio telescope at the South Pole had found evidence that the Big Bang theory of how the universe began checks out.
- New telescopes, which could help scientists understand how the universe evolved, are planned for the South Pole by 2009.

Chapter 21

Scientists Focus on the Arctic

In This Chapter

- What scientists are trying to learn about the Arctic and global climate
- Arctic Ocean life could affect global climate
- A ship is frozen into the ice to help scientists learn how the Arctic works
- A ship full of scientists works to understand how changing climate will affect Arctic Ocean life
- Fresh water will be the next big focus of Arctic research

Concern about Earth's changing climate has been focusing scientific attention to the Arctic since the 1980s. Computer models show that the polar regions, especially the Arctic, should be the first affected by a warming world. Scientists have good reason to think that natural climate cycles are causing at least some of the changes we are seeing. But they don't know why some things, such as a decrease in Arctic sea ice, are happening. In this chapter, we'll look at how scientists are learning what is happening and could happen in the Arctic.

Scientific Interest in the Arctic Grows

As global computer models of the climate grew more sophisticated during the 1980s, scientists began to see how the role of the polar regions in a warming world could be more complicated than merely adding more water to the oceans as ice melted.

Discoveries from ice cores and computer models were indicating that, in a warming world, the polar regions could play a disproportionate role in global climate. Luis Tupas of the National Science Foundation's Office of Polar Programs sums it up: Earth's climate seems to be "a big machine with a small switch." The Arctic Ocean could be a "switch" that can flip global climate to a new mode.

Some Arctic Questions Scientists Are Asking

Scientists agree that the Arctic is showing signs of climate change, Tupas says. What isn't absolutely clear is whether changes in global climate, caused at least in part by humans, are responsible for the many changes seen around the Arctic. Or are Arctic changes merely part of normal up and down swings in climate? The best bet seems to be that it's some combination of the two.

The polar regions are the earth's air conditioners. In effect, scientists want to take off the cover and learn more about how the polar air conditioners work. In this chapter, we look at some of the ways in which scientists are looking at the Arctic air conditioner, and new investigations they would like to make.

Scientists Are Interested in Native Knowledge

In the past, scientists going to any part of the world to study almost any aspect of the oceans or land paid little attention to what the native people had to say. That's changing, especially in the Arctic. "We want to find out what the person on the shore sees," says Luis Tupas of the National Science Foundation. "They can tell us the changes they've noticed. That information is now being used by scientists."

Eskimos Use Science to Save Whale Hunt

We saw in Chapter 15 how the Eskimos around Barrow, Alaska, have been working with scientists since shortly after World War II. In the late 1970s, the Barrow Eskimos' experience with scientists helped save the whale hunt, which is a central part of their culture. In 1977, based on estimates by U.S. government scientists that only 600 to 1,800 bowhead whales remained, the International Whaling Commission ordered a moratorium on Eskimo whale hunting.

Ice Chips

Inupiat Eskimos from northern Alaska have helped scientists determine that bowhead whales can live to be up to 200 years old. Since the 1980s, hunters have found harpoon points made of ivory and stone, which Alaskan whalers haven't used since the 1880s, when they obtained steel harpoons. Scripps Institution of Oceanography and University of Alaska scientists, using changes in amino acids in the lenses of the whales' eyes, came up with the ages up to around 200 years.

The Eskimos tried to convince the government that the scientists had missed a lot of whales in their census, Bill Hess writes in his book, *Gift of the Whale*. "We were trying to explain everything to them, but they wouldn't listen," said the late whaling captain Harry Brower Sr. of Barrow. "I had been trying to tell them, you don't see the whales because they go under the ice and don't show up in the lead. They come up in the young ice. They push their nose up, put a crack in the ice, and breathe .... We've seen it—so many springs!"

Still, the Eskimos, who had learned how to work with scientists, managed to persuade the government to work harder at counting all of the whales, including putting hydrophones in the water to hear whales that scientists, who had stood on land, hadn't seen to count. By 1996, the official estimate of bowhead whales was up to 8,200, and the Inuit not only in Alaska but also in Russia were able to hunt the whales they depend on for food.

Hunters Help Track Seal with Radio

In 2001, native hunters on Little Diomede Island in the Bering Strait captured a ringed seal to which scientists attached a satellite-tracking device. Over the next 7 weeks, they followed the seal as it swam more than 400 miles (700 kilometers) north into the Arctic Ocean, diving deeper than 165 feet (50 meters) from time to time to grab food.

This was the first time a seal had been tracked in the open ocean. Such tracking is a first step in learning exactly how the Arctic's marine mammals live, knowledge that will help determine how a changing climate should affect them.

The project was an offshoot of a National Science Foundation Project to set up an environmental observatory on Little Diomede Island. With the observatory, researchers will regularly collect chemical, biological, and physical data on the nutrient-rich Pacific Ocean water flowing into the Arctic Ocean through the Bering Strait.

The project includes involving Alaska's natives in research, said Gay Sheffield of the Alaska Department of Fish and Game. "The great thing is that you have people sharing information and learning together," Sheffield said. "I was working with men who work with and observe these animals on a daily basis. They are the experts on the animals' local behavior and movements. It was a privilege to be able to unite scientific and traditional knowledge to gain a better understanding of ringed seal life history."

Scientists Need to Understand Heat Budget

Between 1893 and 1896, the Norwegian explorer Fridtjof Nansen let his ship the *Fram* (Amundsen later used it for his South Pole expedition) freeze into the Arctic's ice, inventing a good way to study the Arctic Ocean. Beginning in the 1930s, the Soviet Union learned that scientific stations located on drifting Arctic ice could collect valuable data. The United States followed in the 1950s, but both nations also learned the down side: Drifting ice tends to break up, tossing instruments, structures, and even people into the cold water (if they don't flee in time).

Polar Talk

A **heat budget** is an inventory of all of the heat flowing into and out of a system and the heat stored in the system.

By the middle of the 1990s, climate scientists knew they needed to know more about the Arctic Ocean's *heat budget* during the entire year to have any hope of building computer models that could show how the Arctic fits into the world's changing climate. On a global scale, Earth's heat budget has two components: Energy arriving at the earth from the sun and energy leaving the earth.

The picture, of course, isn't this simple. Throw in ocean water, ice that comes and goes, a season when the sun doesn't rise, another when it doesn't set, and it all becomes fiendishly complicated. Scientists need data collected with a suite of high-tech instruments, 24 hours a day, through all of the seasons. The National Science Foundation decided to invest $19.5 million in freezing a ship into the Arctic Ocean's ice for a year to drift with the ice. Today, big scientific projects need spiffy acronyms; this one was SHEBA for "Surface Heat Budget of the Arctic Ocean."

Canadian Icebreaker Stops Breaking Ice

In October 1997, the Canadian Coast Guard icebreaker *Des Groseilliers* was frozen into the ice about 300 miles (483 kilometers) north of Prudhoe Bay, Alaska. Soon the ice around it sprouted a town of plywood and canvas huts, meteorological towers, automatic buoys, and other instruments. A snow runway allowed Twin Otter airplanes to haul supplies and people in from Alaska.

Ice Chips

More than 170 scientists worked at SHEBA at one time or another, but the largest number there at any one time was 35 to 40 in the spring, and the smallest number was about 15 in the coldest part of winter. The Canadian Coast Guard crew had 16 members, about half the number when the ship is underway. Most of the scientists came and went, but one American stayed six months, and a Canadian biologist never left.

During the year, until the *Des Groseilliers* broke free in October 1998, the winds and currents carried it in a zigzag path of about 1,000 miles (1,609 kilometers) that ended up about 400 miles (644 kilometers) northwest of where it started. The ice, didn't stay in one piece. During an April blizzard, for example, part of the camp moved a quarter-mile (370 meters) away from the ship.

Data from satellites, a high-altitude research airplane, even a U.S. Navy submarine were added to that from the ship. Richard Moritz of the University of Washington, the project director, said: "We've observed the ice, the atmosphere and the ocean over a full annual cycle covering the physical variables in all three systems. We've seen it all: Melting, freezing, heating, cooling, ridges, cracks, leads, melt ponds, and all kinds of different formations of ice and snow." Researchers will be busy the next few years making sense of it.

Ice Chips

During the yearlong SHEBA experiment, the coldest temperature recorded at the station was –44°F (–42°C) on New Year's Eve 1997–1998. The warmest was 33°F (0.8°C) on July 20, 1998.

A Permanent Station Wouldn't Work at North Pole

The North Pole is near the middle of the Arctic Ocean; the part of the ocean that is mostly covered by ice all year. The key word is *mostly*. Leads open up from time to time, even at the North Pole itself, and the ice is always drifting, making a permanent station like the one at the South Pole impossible. From the 1930s into the 1970s, the Soviet Union almost had a North Pole Station with some of its drifting stations coming quite close to the pole.

Unfortunately for scientists, the Soviet Union, which was going broke and unraveling by then, gave up the drifting stations and regular airborne surveys of the area around the North Pole in the early 1980s, just as the science was getting more interesting.

Explorations

Statements like, "The North Pole ice cap is melting," could go back to an August 19, 2000, *New York Times* front page story headlined: "Age-Old Icecap at North Pole Is now Liquid, Scientists Find." It was based on a report from a climate scientist aboard a Russian cruise ship that he had seen open water at the North Pole. The story described the open water as more evidence of global warming affecting climate. Ten days later, *The Times* ran a story headlines: "Open Water at Pole Is Not Surprising, Experts Say." It said, "Although striking and unusual, those reports are not as surprising as suggested in a news article on August 19 in *The New York Times*."

The Arctic Ocean seem to have less ice on average than before the 1980s, and patterns of air pressure, winds, temperatures, and the ocean's water flow have been changing. In the late 1990s, scientists found some evidence that the halocline could be growing weaker. This layer of salty water keeps warmer water below from melting the Arctic's ice. To learn what's going on, scientists need more weather and oceanographic data from the North Pole.

This is why, in April 2000, a team led by James Morison of the University of Washington, set up an automated station to collect and transmit via satellite weather and some basic oceanographic data. The station drifted with the ice for just short of a year, sending data until it drifted into the Greenland Sea. This was the beginning of a 5-year, more than $5 million National Science Foundation project to collect more polar data.

When they returned to the North Pole in April 2001, the researchers set up not only the automated stations to drift with the ice, but anchored a 2.7-mile (4,300-meter) string of oceanographic instruments to the floor of the ocean near the pole. The only previous attempt to do something like this was in 1977, and the mooring lasted only a month. Because the top end of the mooring was about 150 feet (50 meters) under the ice, and couldn't transmit data, it was all stored to be retrieved the next year.

Ice Chips

To set up the North Pole stations in 2000 and 2001, researchers flew from the Canadian Forces Base at Alert, Nunavut, and camped near the pole. In 2002, they used the private Russian ice station set up each spring for tourists to ski or ride hot air balloons the last 60 miles (96 kilometers) to the pole.

When the researchers returned in April 2002, they found the mooring, sent a sound signal to unhook it from its anchor, allowing floats to carry it up under the ice. After cutting a hole in the ice, scuba divers recovered the mooring, the instruments, and data recorder. They replaced it with a similar mooring to be retrieved the next spring, which they did in April 2003.

Scientists Sail in Search for Answers

The July sun never sets on the Arctic Ocean, which helped the 38 scientists aboard the Coast Guard icebreaker *Healy* in their around-the-clock probes of the ocean's life. They didn't need lights to lower their collection devices into the cold ocean. Some of these were almost as simple as the net a child would use to catch minnows; others were high tech. Water and the things in it were sampled from the top to the bottom of the ocean. They also hauled aboard ocean-bottom mud, and the creatures crawling in it.

The *Healy*'s cruise in the summer of 2002 was a part of the multi-year Western Arctic Shelf-Basin Interactions project. The National Science Foundation and the Office of Naval Research are supporting it as part of the long-term goal of leaning how the Arctic Ocean's life could affect, and be affected by, changing climate.

Explorations

One of the basic devices oceanographers use is a "conductivity-temperature-depth (CTD) rosette" A CTD is lowered into the water—the one used on the *Healy* in 2002 could go down 10,000 feet (3,000 meters). It measures the conductivity of the water for precise data on how much salt is in it. The CTD also includes additional instruments that measure how deep the light green plants need penetrates into the water, and the amounts of substances such as dissolved oxygen, and nitrates at various depths. As the name says, it measures the temperature, and it measures the depth. (Otherwise the other measurements wouldn't be very useful.) The tanks arrayed around the "rosette" are opened to take in water from whatever depths the scientists are interested in.

A conductivity-temperature-depth (CTD) rosette is lowered into the Arctic Ocean from the Coast Guard icebreaker Healy.

(Photo by Jack Williams)

The "shelf" in the project's name is the continental shelf. This is the area where the water is no deeper than 350 feet (100 meters) as it slopes away from Alaska's northern coast. The "basin" is the deeper part of the Arctic Ocean. In Chapter 6, we looked at how the Arctic Ocean's food web begins with plankton. It turns out that a great deal of this plankton is found in the shallow waters of shelves, but the food web it supports extends into deep water. Exactly how this works is one of the things the Shelf-Basin researchers are trying to untangle.

How could Arctic Ocean plankton, or for that matter, the largest Arctic Ocean whale, have any effect on the global climate? Although no particular piece of plankton or an individual whale could affect the world's climate, they are part of the *biosphere*, and the biosphere very much affects the climate.

Polar Talk

The part of the earth, including the oceans and the atmosphere, where life is found is called the **biosphere.** Used by climate scientists to refer to living things as they are affected by the atmosphere and in turn affect the atmosphere.

Polar Talk

Gases that absorb heat being given off by the earth are called **greenhouse gases** because they then radiate energy in all directions, keeping the earth warmer than it otherwise would be. This warming is called the "greenhouse effect."

To see how this works, let's look at the end of the Arctic Ocean food web. As we saw in Chapter 6, the carbon created by plankton works its way through the food web. When plants and animals die, they settle to the bottom of the ocean and their carbon doesn't get into the atmosphere. Other animals eat some of the carbon in plants and animals. As animals turn what they eat into energy, some of the carbon is exhaled into the ocean (by fish) or the air by seals, whales, polar bears, and people.

Carbon dioxide is one the *greenhouse gases* that keeps Earth warm enough for life as we know it to exist, but as more is added, it tends to turn up Earth's thermostat. One of the questions scientists are trying to answer: Will a changing climate make the Arctic Ocean a source of atmospheric carbon dioxide going into the air, or will it remain the "sink" it is now, which locks up carbon on the ocean's bottom?

Today's Data Will Help Model the Future

The scientists on the 2002 cruise, and on other Arctic cruises that are planned as part of the Shelf-Basin project, are after detailed, nuts and bolts information on how the Arctic Ocean's water and currents and its life ranging from microscopic bacteria to whales and polar bears all work together to create a unique environment and how this is changing as the Arctic warms.

Explorations

The National Science Foundation and the U.S. Coast Guard worked together to design the icebreaker *Healy* not only to break ice, but also to be a scientific research vessel. It went into service in August 2000. The 420-foot (128-meter) ship has 7 laboratories with a total of 5,000 square feet (455 square meters) of space. It is highly automated with computer systems designed for science. It can break 4.5 feet (1.4 meters) of ice at a steady speed of 3 mph (5 kph). It can operate in temperatures as low as −50°F (−45.5°C). It's named for Captain Michael Healy, the son of a former slave, and noted navigator of the Bering Sea and Alaskan Arctic as commander of the U.S. *Revenue Cutter Bear* from 1886 to 1895.

Fresh Water Will Be a Focus of Arctic Science

Ice and water flowing from the Arctic Ocean into the Atlantic Ocean around Greenland seems to be one of the key drivers of a global system of ocean currents. Scientists have good reason to think that changes in these currents could have global climate consequence. Although the Arctic Ocean, like all oceans, is salty, it's obvious that the amount of fresh water flowing into the ocean, or falling on it as snow, is an important part of the system. This is why the next large Arctic Ocean science focus will be Arctic fresh water, in all of its forms. Luis Tupas of the National Science Foundation says the Foundation wants scientists to look at fresh water in the ocean, in the rivers that flow into the Arctic Ocean, the snow that falls around the ocean, what's happening with permafrost, and changes in all of them.

Knowing more about the Arctic Ocean, the atmosphere, the fresh water flowing into the ocean, and the oceans and atmosphere elsewhere could help us make sure that a "climate switch" isn't thrown without our being ready for it.

The Least You Need to Know

- Climate warming could change the Arctic Ocean and the land around it, affecting the rest of the world.
- Data from Arctic Ocean studies could improve computer models of global climate.
- An icebreaker spent a year frozen into Arctic ice allowing researchers to gather data on heat movements in and above the ocean and ice.
- Researchers want to learn how Arctic Ocean life is involved in a changing climate.
- Researchers are beginning an intensive study of fresh water that comes into the Arctic Ocean.

Chapter 22

The Mysteries of Antarctic Ocean Life

In This Chapter

- The Southern Ocean is full of life
- Scientists are trying to learn how the ozone hole affects Southern Ocean life
- Winter research uncovers how tiny creatures survive
- An unmanned submarine finds krill under the ice
- Penguins and ice waltz to a complex tune

Despite its harsh environment, the Southern Ocean harbors a great deal of life from phytoplankton at the bottom of the food web up through a variety of birds including penguins, and marine mammals such as whales. Concerns about the potential effects of ozone depletion and climate change drive much of the scientific research in and around the Southern Ocean. In this chapter, we look at some of this research ranging from the bottom of the food web to penguins, whose lives are dominated by sea ice.

A Simple Food Web Survives a Harsh World

As we saw in Chapter 7, sea ice provides a winter home for algae that blooms when the sun finally returns in the spring. The algae is at the bottom of a food web that's about as simple as you can find. The ice-based algae feed a community of zooplankton, bacteria, and almost-microscopic creatures and tons and tons of krill—shrimplike creatures that fish, seals, penguins, and whales eat. A great deal of Southern Ocean biological research has the ultimate aim of discovering how a changing climate could affect life there because that life is so dependent on global ocean currents, which are a key part of the global climate system.

The Ozone Hole Prompts Biological Studies

In 1985, scientists discovered that the amount of ozone in the *stratosphere* over Antarctica thins dramatically each spring and then recovers during the summer. Scientists gave this thinning the name "ozone hole." In Chapter 25, we'll look at how it was discovered, what causes it, and why scientists have good reason to think it will close during the twenty-first century.

The Antarctic ozone hole raised concerns because some scientists thought it could be a forerunner of what could happen around the world. But the most immediate concern was the hole's potential to damage Southern Ocean life and Antarctica's few plants. Stratospheric ozone absorbs a great deal of the *ultraviolet light* coming from the sun, especially the most harmful wavelengths.

Crevasse Caution

The "ozone hole" is not really a hole, but a decrease in the amount of ozone over Antarctica, and in some years over the Arctic, for part of the year. It is certainly not like a hole in a balloon that will let the earth's atmospheric gases escape.

Polar Talk

The **stratosphere** is the layer of the atmosphere that begins about 4 miles (6 kilometers) above the earth in the polar regions, and about 10 miles (16 kilometers) above the tropics. The top of the stratosphere is about 31 miles (50 kilometers) up all over the earth.

Ultraviolet light is the invisible form of light with frequencies just higher than those of violet light. Ultraviolet energy causes sunburn and harms living matter in other ways.

While humans can defend themselves against ultraviolet light, plants and animals are stuck with whatever defenses they've evolved over eons. The ozone hole could threaten phytoplankton at the base of the Southern Ocean's food web. No one knew what the additional ultraviolet energy reaching the Southern Ocean was already doing, what the long-term effects would be, or what more ozone thinning would do. Scientists needed to find out.

Life Seems to Be Coping with Ozone Loss

The good news is that so far, scientists have not detected any signs of disastrous changes in Southern Ocean life. They have seen some effects, however. The best estimate is that the added ultraviolet light has reduced the amount of phytoplankton in the Southern Ocean by 6 to 15 percent, says Deneb Karentz of the National Science Foundation.

Less phytoplankton affects the entire food web, because that's what other creatures, including krill, which fish, birds, seals, and even whales eat. "Is 15 percent enough to cause a decline in krill? "We don't know," Karentz says. "But it's very difficult to exactly analyze the impact of the ozone hole on the Antarctic."

Ice Chips

The Antarctic ozone hole is of little direct threat to humans because it occurs where people don't sunbathe. People in Antarctica were already using sun block and sunglasses because of the glare from the snow and ice. The very southern tip of South America is the only inhabited area affected, and then only briefly. Early 1990s reports of ultraviolet light blinding sheep in southern South America turned out to be bogus.

Before the discovery of the ozone hole, no one had studied much of the life that the additional ultraviolet radiation would be likely to harm. "Unfortunately, there are a lot of other factors that affect the populations," Karentz says. These include year-to-year weather and sea ice changes. "Being able to separate the effects of ultraviolet from the others is a problem. There have been some changes, but the system is resilient enough, things will survive."

Polar Talk

DNA is the abbreviation for deoxyribonucleic acid, which is a thin, chainlike molecule found in every living cell. It directs the formation, growth, and reproduction of cells and organisms. Short sections of DNA called genes determine heredity; that is, the passing on of characteristics in living things.

All Life Defends Against Ultraviolet Light

One reason that the system is resilient is that "all forms of life from bacteria to humans have evolved to deal with ultraviolet light. All have ways to protect themselves and to repair damage," Karentz says. In her studies, she has found a form of natural "sunscreen" in phytoplankton that is common in marine algae and invertebrates, and even in fish eyes. Also, life from algae to humans has evolved *DNA* repair mechanisms. The question is: How much ultraviolet light is needed to overcome these defenses, especially those of the most vulnerable forms of life?

Scientists Probe What Happens After Dark

The Southern Ocean's sensitivity to climate change and its relatively simple food web, make it a perfect part of a worldwide study of how global climate change may affect life in the oceans. The study, called GLOBEC for GLOBal ocean ECosystems dynamics program, involves scientists from around the world. Unlike the scientists who are studying effects of the ozone hole, U.S. GLOBEC researchers want to learn what happens to Southern Ocean life all year, during the dark polar winter as well as in summer.

Ice Chips

"As winds reached up to 69 mph (111 kph) through the Drake Passage … scientists experienced firsthand how life in the ocean depends on the movements of the sea," Kristin Cobb, a science writer on one of the cruises, wrote. "Cycling between weightlessness and descent, the sensation was much like being on an endless roller coaster. Once nausea hit, time seemed to pass very slowly."

Until a 1986 winter cruise by the German research vessel *Polarstern*, scientists knew little about ocean life in the dark. For instance, they assumed that krill hibernate on the ocean floor. Although scientists today know that's not the case, exactly how krill and other Southern Ocean life make it through winter is far from understood. But they want to know because, "what happens in the winter determines how productive the ecosystem is," said Eileen Hofmann of Old Dominion University in Norfolk, Virginia, a U.S. GLOBEC researcher.

The *Lawrence M. Gould* and the *Nathaniel B. Palmer*, the National Science Foundation's two research icebreakers, made five cruises to the west side of the Antarctic Peninsula from March through September 2001, and five more in 2002. One group of scientists attached radio-tracking equipment to Adelie penguins and crabeater seals to find out where krill and other prey go in the winter. Hungry penguins or seals would be sure to find the food they need for survival. The researchers discovered that the seals dived deeper and longer than scientists had seen during similar studies in the summer. This would fit with the seals going after krill that move up and down in the water in different seasons.

One finding, which fits with others about Southern Ocean life, is that krill reduce their metabolism in winter, Kendra Daly of the University of South Florida, said, adding: Krill eat bacteria and algae on the bottom of sea ice. "It's kind of like spinach in your freezer. Because winter food supplies are low, these "algal Popsicles" are a key food source. Some krill will starve if sea ice doesn't form, or forms late in the winter.

Unmanned Sub Finds Where Krill Hang Out

Krill congregate under sea ice from around ½ mile to 8 miles (1 to 13 kilometers) back from the ice edge, British Antarctic Survey researchers discovered when they sent an automated, unmanned submarine under the ice to look around. The finding was the first consistent comparison of the numbers of krill under the ice and in open water. Krill are about five times more abundant under the ice than in the open.

Autosub leader Andrew Brierley said the krill discovery "is a major new insight because it shows that it is the ice edge, rather than sea ice generally, that is important for krill. Large reductions in ice area, perhaps following melting due to regional climatic warming, might not be so harmful to krill populations because, in summer at least, krill are not widely distributed under ice."

Ice Chips

In addition to studying krill, the British Autosub-2 made some measurements of ice thickness. The U.S. National Science Foundation expects to use similar unmanned subs to study ice thickness and to collect other data in polar oceans, says Luis Tupas of the NSF. For some studies, subs would be better than icebreakers because they wouldn't disturb the ice.

Researchers Use Technology to Probe Penguins

As we saw in Chapter 9, Antarctic penguins live in an unbelievably harsh environment. Today's penguin researchers are trying to learn how penguins manage to pull this off and how past climate changes have affected them, with the aim of getting a handle on the effects of a warmer world. As anyone who's looked at a penguin *rookery* could tell you, picking out one penguin from another is virtually impossible. Yet being able to follow particular birds can be quite fruitful. Enter technology.

Your veterinarian can implant an identification chip about the size of a grain of rice in your pet. If Fluffy or Fido wanders off and ends up in an animal shelter that scans animals, you'll get a call to come pick up your lost pet. Penguin researchers use similar chips to track individual birds.

Polar Talk

A **rookery** is an area where large numbers of birds or mammals, such as penguins or seals, come together to breed or live.

One study involves putting maybe 50 breeding pairs in a fenced area with only one opening, forcing birds to walk on a scale as they come and go. Each time a bird leaves to feed, or returns with a full belly, its weight is recorded. Researchers use this data to learn things such as how long birds spend looking for food, how much food they bring back to their chicks, and if the guy and the gal penguins have different feeding strategies.

Gerald Kooyman of the Scripps Institution of Oceanography and his associates, using satellite tracking technology, have found that emperor penguins range over almost all the Ross Sea, which is about as large as Texas, disproving the old idea that they were homebodies that hunted close to home.

Kooyman was the first to use time-depth recorders to see what diving animals, such as penguins, do when they disappear from our sight. The device, which is glued to a penguin's feathers, records the time and how deep the bird goes. It includes a radio transmitter that helps researchers find the bird when it returns from feeding. Devices that researchers don't recover fall off with the old feathers the next time the bird molts. Kooyman's group has recorded one emperor penguin that dove 1,640 feet (500 meters) down.

Adelies Are Picky About Where They Live

David Ainley, an Adelie penguin researcher, notes on his website that Adelie penguins have been around as a species about three million years, just about the same amount of time that modern humans, *Homo sapiens*, have been roaming the earth. "Both species have existed during a period of unstable climates, marked by a series of dramatic shifts between glacial and interglacial periods. Climate change is not new to either and, dear reader, mark the following words carefully: both moved in response."

While humans are scattered all over the world, Adelie penguins live around Antarctica and nearby islands. Ainley estimates that the 161 colonies have around 2.4 million nesting pairs of birds. As the world has warmed since the height of the last Ice Age, around 19,000 years ago, Adelie penguins have been moving south and onto the Antarctic continent, which was entirely covered by ice during the Ice Age, and, thus, had no penguins.

Adelies live in areas where there is sea ice just in the way forest birds nest where there are trees, Ainley says. This is probably why the numbers of Adelie penguins are decreasing around the Antarctic Peninsula where the temperatures are getting warmer and winter sea ice is decreasing.

But if there is too much sea ice between their colonies and open water, Adelie penguins go elsewhere. Penguins have evolved unusually streamlined bodies that allow them to swim fast and for long distances while conserving energy. But penguins are awkward out of the water. If getting to and from their colony and open water where they find food begins to take too long, their nests will fail, and the birds will move elsewhere.

Ice Chips

Chinstrap penguins are very much like Adelie penguins, David Ainley says. In fact, some scientists think the two are a super species. Both feed mainly on krill. Yet unlike the Adelies, which live only where there is sea ice, chinstraps prefer the open ocean. As the Peninsula warms and the Adelies move south, chinstrap penguins are spreading south along the Peninsula.

Ice Is Nice, but Enough Is Enough

While both Adelie and emperor penguins, the two species that nest on the main part of Antarctica, live with ice, they had too much of it around Ross Island during the 2001–2002 and 2002–2003 summers. Two large icebergs that moved to the northern end of Ross Island acted like breakwaters, keeping ocean waves, which normally help break up sea ice in the spring, out of McMurdo Sound.

One of the bergs also pushed broken ice onto the shore around the big Cape Crozier emperor colony, restricting access to open water, and almost wiping out the colony. The colony's failure is helping scientists learn more about how penguins respond to changes in their environment; knowledge that might hold clues to the future of not only Emperor penguins, but other Southern Ocean creatures.

The extra sea ice in McMurdo Sound gave Adelie penguins at the Cape Royds colony on Ross Island—the world's southernmost penguin colony—too far to walk. "If sea ice does continue to form each winter as it did (in 2001–2002), then in about 10 years the Cape Royds colony will be history," Ainley says.

But this is just one colony. Antarctica's Adelies are not threatened; the millions of birds elsewhere around the continent were not affected by what happened on Ross Island. Ainley says the failure of the Cape Royds colony would be merely a "'burp' in a very successful and rapid re-colonization of Antarctica" by Adelies since the end of the last Ice Age.

The Least You Need to Know

- Global climate change could affect life in the Southern Ocean around Antarctica.
- Added ultraviolet light that reaches the Antarctic through the ozone hole seems to be doing little damage.
- Researchers made 10 cruises to study life all year in the ocean off the Antarctic Peninsula.
- An unmanned submarine discovered that krill cluster under the edge of sea ice.
- Penguin researchers today use high-tech devices to gather detailed data on how the birds live.
- Large icebergs and unusual amounts of sea ice have killed many penguins around Ross Island, but Antarctica's other penguins were not affected.

Chapter 23

The Mystery of the Cooling Continent: Antarctica

In This Chapter

- Antarctica was once warm
- Antarctica drifts to the South Pole
- The magnetic poles flip
- Earth stashes secrets on Antarctica's sea floor
- Meteorites strewn on Antarctica's ice fields, tell of far away worlds

Antarctica is a strange and exotic land. It holds secrets of a mysterious past that illuminates Earth's turbulent, molten-iron-driven history. Alien rocks rain upon Antarctica's icy fields and hold secrets of far-away worlds. In this chapter, we'll see how scientists have solved some of these mysteries, but many remain.

Antarctica Began as Part of the Super Continent

Since Heroic Age explorers first found coal and fossils early in the twentieth century, scientists have known that Antarctica was once warm. During the last decades of the twentieth century, researchers using fossils and other methods have put together the story of how a once-warm Antarctica became The Ice we know today.

Antarctica's story begins 250 million years ago when all the continents existed as one stretching form pole to pole, called *Pangaea*, with the proto Antarctica in the south. Here, four-foot long amphibians slid into ponds and eyed dragonflies as big as parrots flitting overhead.

Polar Talk

Pangaea was the super continent that broke up around 200 million years ago to begin forming the present landmasses.

The earth's **mantle** is the layer of Earth between the crust and the core.

The land was turbulent. Earth's *mantle* always seethes like thick soup in a pot; great convection currents bulge against the crust. Sometimes the crust cracks between tectonic plates. Pangaea felt such forces. Faults zigzagged through northern mountains. A long narrow sea poked into Pangaea's east coast. Rifts crossed the land. Lava oozed up from the mantle and spilled into the rifts. Barely formed, the giant continent broke into smaller ones.

Polar Talk

Gondwanaland was the proto continent of the Southern Hemisphere that broke up into India, Australia, Antarctica, Africa, and South America.

By 200 million years ago, Pangaea had split into two parts: a northern Laurasia and a southern *Gondwanaland*, with Antarctica forming Gondwanaland's southland. At this time, warm shallow seas lapped Gondwanaland's southern shores. It had a Florida-warm, moist climate, as did the rest of the earth.

Polar Talk

A **pterosaur** is an extinct flying reptile.

By 175 million years ago, the Antarctica part of Gondwanaland was centered at 65 degrees south latitude—much farther north than now. Large and small meat-eating dinosaurs, long-necked plant-eating dinosaurs with small heads, and beaver-size, reptile-like mammals walked the land. *Pterosaurs* flew in the skies. The mammal-like reptiles (called therapsids) left two-inch tracks in the mud.

How Antarctica Turned from Warm to Cold

Over the epochs, Gondwanaland fragmented into pieces that formed the continents we know: Africa, South America, India, Australia, and Antarctica. The pieces drifted apart as slowly as fingernails lengthen—an inch a year. Finally (160 million years later), Antarctica was in its present position—totally isolated and ringed by frigid sea currents. The theory of plate tectonics explains how such an event could occur: The outer hard crust of Earth that seems so fixed and continuous to us, actually consists of a dozen or so distinct, hard plates that drift individually on hot, deformable rock like huge ice floes on a polar sea. An unequal distribution of heat within Earth moves the plates much like clumps of floating parsley drift about the surface of simmering soup.

Explorations

In 1912, Alfred Wegener, a German meteorologist, first developed the plate tectonics theory. When he was young, Wegener wanted to be a polar explorer. After earning a Ph.D. in astronomy, he turned to the then-new science of meteorology because it offered a chance to work with expeditions in Greenland. Wegener went on four of them. His scientific work ranged far beyond meteorology. Wegener's "continental drift" theory wasn't generally accepted until new research methods unearthed overwhelming evidence in the 1950s and 1960s that the continents move on plates; thus the name "plate tectonics." Wegener didn't live to see the findings that proved his theory, including those from Antarctica. He died, shortly after his fiftieth birthday, in November 1930, when he became lost in a blizzard while taking supplies to a weather research station on Greenland's Ice Cap.

When the water between Australia and Antarctica opened about 40 million years ago, Antarctica was isolated with the Southern Ocean free to swirl unhindered around the continent that surrounded the South Pole. Warmer water could no longer reach and heat it. As the climate cooled over the next five million years, ice sheets formed. Animals, now trapped by the sea, couldn't escape. Neither could they adapt to the extreme cold. Species went extinct like stars winking out as night clouds gather.

More ice accumulated as the temperatures continued to fall. Ice fields merged and thickened until they almost buried the 13,000-foot (4,000-meter) Transantarctic Mountains. The sheets surged down to the coast and into the sea. By 10 million years ago, ice capped Antarctica as it does now.

Lava Lake Simmers Atop an Antarctic Mountain

Earth still shifts continents over its hot mantle as it did when Gondwanaland fragmented. The Atlantic Ocean, for example, widens an inch a year, splitting North America from Europe. For the last 40 million years, a triple junction in the Indian Ocean (off the coast of South Africa) has separated three plates: Africa, Antarctica, and Indo-Australia.

As a result, Antarctica has a live volcano—Mt. Erebus—with a simmering lava lake covering a small part of its crater. Small gas bubbles burst on the lake's surface and fizz like a freshly opened Coke bottle. Extreme subterranean pressures and heat force gas into a lava solution. When the lava oozes to the surface, the gas bubbles out of solution, occasionally hurling a *volcanic bomb* from the lake. Bombs, some the size of refrigerators, litter the crater floor, but Erebus hasn't erupted since 1841, when James Clark Ross happened to sail into the sea now named after him.

> **Polar Talk**
>
> A **volcanic bomb** is a semi-molten, streamlined fragment of compressed ash and pumice, which is ejected from a volcano.

Magnetic Poles Shift and Switch

Earth's seething mantle not only moved the continent of Antarctica to the South Pole, it continues to shift the North and South magnetic poles. The South Magnetic Pole is now more than 550 miles from where Shackleton's team found it in 1909. It's now 150 miles off the Antarctic Coast in the Southern Ocean.

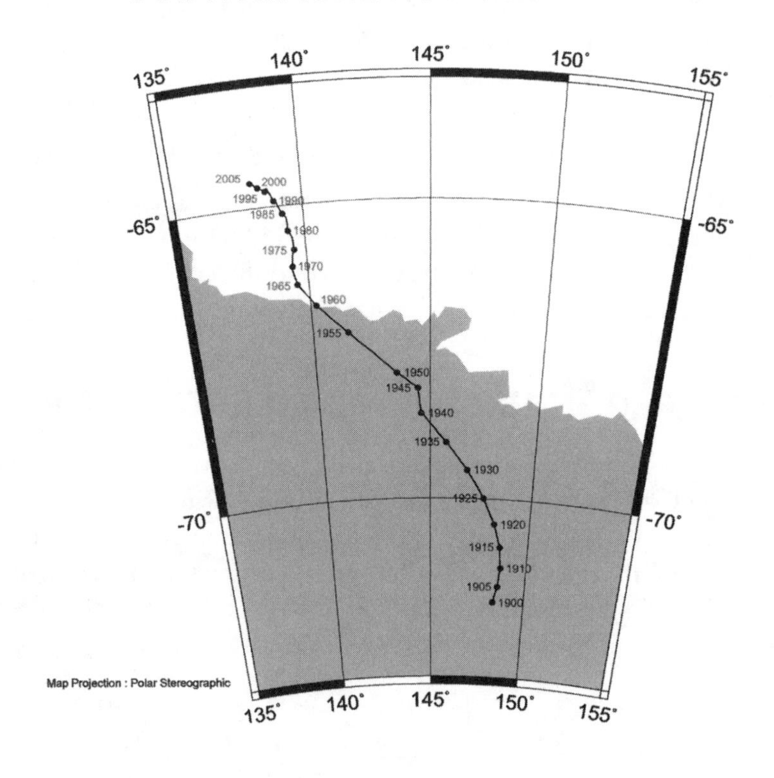

Locations of the South Magnetic Pole.

(U.S. Geological Survey)

The magnetic poles have not only moved around, they have flipped (north becoming south and visa versa) more than 20 times in the past 5 million years. By the way, magnetic poles are not necessarily 180 degrees apart.

A flow of electric current deep inside Earth maintains the *magnetic field* that has existed for at least three billion years. The core of the earth is made of iron-*alloy*—a solid ball surrounded by a molten shell. The cooling earth solidifies some of the molten iron from the alloy in the shell. This change from liquid to solid releases heat. Moreover, the heavier, cooler alloy sinks and pushes up the lighter buoyant alloy. This circulation creates moving electrical charges, generating a magnetic field.

The combination creates a great roiling movement in the rest of the molten core. The Coriolis force caused by Earth's rotation, twists and shears the flow into a helical pattern. The motion of an electrically charged fluid is an electric current. Earth's core functions, in this fashion, somewhat like an *electromagnet*.

Polar Talk

A **magnetic field** is a force field in the region around a magnet or an electric current, characterized by lines of magnetic force connecting the two poles of the magnet or surrounding the electric-current flow. An **alloy** is a homogeneous mixture of two or more metals, the atoms of one replacing or occupying positions between the atoms of the other. Brass is an alloy of copper and zinc. An **electromagnet** consists of a coil of insulated wire wrapped around a soft iron core that is magnetized only when current flows through the wire.

The current causes an electromagnetic field about Earth similar to that of a nail wrapped with a wire connected to a battery. The twisting flow pattern of the molten alloy is slightly unstable and sometimes this generates a new magnetic field oriented in the opposite direction. When this happens, our poles switch. The last pole flip occurred about 800,000 years ago.

Antarctica Hoards History on Ice, Under the Ocean

Over the past 35 million years, since ice first buried Antarctica, the continent has changed little. It hoards history on ice fields and under seas. It hoards not just Earth's history. Antarctica captures and preserves rocks raining in from the asteroids, comets, the Moon, and Mars. From this treasure-trove, scientists have been able to piece together the history of our solar system.

Ocean-Bottom Sediments Record the Distant Past

Over the eons, dead plants and animals eventually wash out to sea. Mud covers the debris and it becomes part of the sediment record.

Earth's magnetic poles switch at random intervals, from tens of thousands of years to more than a million years apart—an average of once each 200,000 years. Iron-rich lava solidifies, forming permanent magnets aligned with the earth's magnetic field; thus recording the new orientation.

Explorations

Iron-rich rock has a peculiar property: If you heat it above 1,076°F (580°C), it loses its magnetism. When it cools below this temperature, the rock is re-magnetized in the direction of the earth's existing magnetic field. It becomes a magnet with poles aligned with the poles of the earth when it cooled. The neat thing about this is: The magnetic field of the rock, once cooled, stays frozen in this orientation.

Geologists, who know the language of sediment, are probing the sea bottom just offshore from Cape Roberts on McMurdo Sound. Investigators are drilling into the sea floor for cores—from sediment hardened into rock—that document history for at least the past 65 million years. The investigators aim to reconstruct histories: glacial, climate, the uplifting of the Transantarctic Mountains, and how Gondwanaland broke up.

Nine miles (15 kilometers) offshore in McMurdo Sound, a 50-ton drill rig looms like a rocket launch pad on 6-foot (2-meter) thick sea ice. Its bit reaches down into the sea and finds bottom 330 yards (300 meters) below. The bit bites into rock and drills through 15 meters of uninteresting stone. Then it hits millions-of-years-old rock sediment, extracting intact cores. From these cores, the geologists learned that the sea level changed many times between 34 and 17.5 million years ago. The ice cap depth and extent fluctuated strongly and frequently—but didn't totally disappear from Antarctica during that time. Sediments and fossils indicate how the climate changed over the past 34 million years: cool-temperate (34 million), cool (25 million years ago), to the present polar climate.

Rocks from Outer Space Bombard Antarctica

On January 22, 2000, in the Elephant Moraine about 160 miles northwest of the U.S. McMurdo Station, a sporty, white, Volkswagen Beetle–size robot called Nomad crawls over the ice on four huge metal-studded snow tires. A glittering blue-white glare mesmerizes human searchers but not Nomad. Back and forth it goes, its black antenna sticking up for human communication. But no one bothers it. It's on its own to prove itself.

Nomad finds a rock—it's the eighth so far today after tooling along 1,050 feet (320 meters) The first seven were Earth rocks. *Is this extraterrestrial?* It asks itself once more and goes to work. It sticks its manipulator arm with its visible-to-near-infrared spectrometer within a centimeter of the rock face and analyzes the light reflected from the rock to determine its composition. It compares the spectrum with a database for the right shape, color, and other *meteorite* characteristics. Bingo! It's found a meteorite! Its first ever.

Polar Talk

A **meteorite** is a stony or metallic mass of matter that has fallen to the earth's surface from space.

That day on the ice field, the robot, built by Carnegie Mellon University's Robotics Institute in Pittsburgh, examined more than 100 rocks. It found and correctly classified five meteorites out of seven. One that it thought was an Earth rock was a meteorite so rare that the robot didn't have the data in its base to correctly identify it. Robot error? Uh, uh. Nomad might think otherwise.

Antarctica is the best place on Earth to search for space rocks. A "falling star" hits the ice and stays there—visible, a dark object on white ice—in pristine condition for thousands of years. After it cools, the ice doesn't melt about it and nothing erodes it or the nearby surface. Glaciers, moving slowly over the millennia, concentrate the meteorites at certain sites, like the Elephant Moraine.

Twenty tons of meteorites and cosmic dust speckle Earth each year. Since 1976, the National Science Foundation has funded a human search that, combined with Japanese and European efforts, has found 20,000 meteorites sprinkled on Antarctica's ice. The largest, weighing 300 pounds, was part of an 840-pound meteorite that broke into 40 scattered pieces.

Most rocks hitting Earth come from the *asteroid* belt between Mars and Jupiter. When the solar system formed, some Mars-size bodies condensed from the debris in the asteroid belt. These planets collided with each other, broke apart, and scattered their pieces—small rocky objects that we know as "asteroids"—throughout the belt. They constitute the vast majority of meteorites.

Polar Talk

Asteroids are small celestial bodies that revolve around the sun, with orbits lying chiefly between Mars and Jupiter and characteristic diameters between a few and several hundred miles.

Thirty-four meteorites found on Antarctica came from the Moon and one potato-size rock from Mars. Some large body probably hit the Moon and splattered the material into an Earth-crossing orbit. Earth's gravity brought the lunar rocks to Antarctica. A comet or asteroid may have blasted the Martian rock loose 16 million years ago. The debris floated through the solar system, taking its own sweet time, and arrived in Antarctica 13,000 years ago.

In 1984, searchers found the Mars rock nestled in the Allan Hills north of the Dry Valleys. First scientists thought it contained fossils from Mars—evidence of life. Now they think not.

But either way, meteorites hoard extraterrestrial history. It's cheaper to pick up Martian rocks on Earth than on Mars, especially with whizzes like Nomad around.

The Least You Need to Know

- Antarctica was warm for 200 million years.
- Antarctica wasn't always at the South Pole but drifted there over the eons.
- Ice capped the continent 10 million years ago.
- Earth's magnetic poles switch every 200,000 years on the average—north becoming south and visa versa.
- Cores of sediment from the ocean bottom tell the story of Antarctica's past.
- Meteorites found on Antarctica help scientists learn about the solar system.

Chapter 24

Life in Extreme Environments

In This Chapter

- At times, Antarctica is like another planet
- A dull-looking lake turns out to harbor ancient life
- Space-exploration technology will help discover what's under a lake's ice
- Lakes are discovered deep under Antarctica's ice
- Discoveries show life exists in places never thought possible

At first glance, all of Antarctica away from the coast appears to be lifeless, but microbes, algae, plants, and tiny animals live there on the very edge of existence. This makes it of great interest to those who study life in extreme environments, including those who wonder what kinds of life might exist elsewhere in the universe. In this chapter, we will look at some of this life and how scientists are increasing our understanding of life on the edge.

How to Visit Europa or Mars Without a Rocket

Sometimes in Antarctica you can almost forget you are on today's Earth. After two hours of seeing nothing but ice outside your airplane window, it's easy to imagine that you're flying over Jupiter's ice-covered moon Europa. In the ice-free Dry Valleys, you might say: "This looks like spacecraft photos of Mars."

The resemblances go beyond appearances. Europa is covered by ice, and Mars is dry. The biggest difference is that Antarctica has plenty of water, even if it's almost all ice. A basic principle of biology—whether on Earth or a far away planet—seems to be that life needs liquid water.

Ice Chips

While writing his Mars trilogy, Kim Stanley Robinson read about Antarctica to understand what Mars was like in its early days, he told the *Antarctic Sun*. "I thought parts of the Dry Valleys looked very much like photos of Mars. The field stations and even McMurdo reminded me of what I thought early Mars stations might be like." Robinson's novel *Antarctica* is set there in the middle of the twenty-first century.

Spacecraft have found that Jupiter's moon Europa could have an ocean—unfrozen water—under its thick ice. Warmth from inside Europa, could keep the water from freezing. Where there's water, life could exist.

Polar Talk

Astrobiology is the multidisciplinary study of the possible origin, evolution, distribution, and future of life, and the likely limits on life elsewhere in the universe. Astrobiology uses methods of the biological and physical sciences.

Mars appears to have been more hospitable to life far in the past. Today, Mars has polar ice caps made of carbon dioxide ice ("dry ice") and some water ice. There's evidence that in the past, when Mars had a thicker atmosphere, it could have had ponds, lakes, and maybe even oceans. Maybe life could be hanging on there.

Scientists in the new field of *astrobiology* say the geology on other planets and their moons must follow the same rules as on Earth. But life in such places could be different, just as life in Antarctica's ice is different from elsewhere on earth.

Scientists Drill into a Frozen Lake

From the 1950s, when people first saw it, until the 1990s, scientists ignored Lake Vida in Antarctica's Dry Valleys, because "they thought it was just a big block of ice," says Peter Doran of the University of Illinois at Chicago. But in the early 1990s, Doran and some colleagues took a look at the lake with ground-penetrating radar. To their surprise, the radar showed a pool of water at the bottom, underneath 62 feet (19 meters) of ice.

This looked interesting. In 1996, Doran and others, including John Priscu of Montana State University, returned with ice coring equipment. The researchers stopped drilling at 39 feet (11.7 meters) to make sure they didn't contaminate the water below.

Scientist Revives Ancient Life from Ice

When John Priscu thawed ice from the bottom of the core in his Lab, some microbes it in revived. This pushes back the known boundary of how tough life can be. "What does it take to extinguish an ecosystem when you turn the heat down?" asks Doran. "When does life say, forget it, I can't live at this temperature?"

The ice was about 10°F (–23°C) and has probably been that cold since it first froze around 2,800 years ago—before the rise of ancient Greece. Scientists had known that bacteria frozen in 30°F (–1°C) ice could be revived. Doran says: "Vida changes our ideas of what it takes to extinguish life on Mars. This is important for Mars, but also for the time on Earth around 550 million years ago when it's believed the oceans all over the world were frozen." Life could have survived this *snowball Earth*.

Ice Chips

The 3-mile (5-kilometer) long Lake Vida is one of the largest in the Dry Valleys. This area receives less than 4 inches (10 centimeters) of snow a year. The yearly average temperature is around –22°F (–30°C).

Polar Talk

Snowball Earth is the theory that ice covered all, or almost all of the earth, during four super ice ages, each millions of years long, from around 600 to maybe 800 million years ago. Geologists disagree about how extensive Earth's ice cover was during this time and whether thick ice ever covered even tropical oceans.

Lake Vida Microbes Are Old, Not Exotic

Doran says the microbes appear to be mostly cyanobacteria (sometimes called blue-green algae) similar to the communities found in the shallow zones of the other Dry Valley lakes, and in the ice covering some of these lakes. Lake Vida is what's known as a "sealed lake" because the ice blocks new water from actually flowing into the deep part of the lake, which is why the ancient microbes were found in the ice.

Crevasse Caution

You might worry about ancient microbes such as from Lake Vida "escaping" causing disease. Peter Doran says that's not a concern because the microbes "are in water not in air. So even if they could make anyone ill, which is a stretch, someone would have to drink the brine for that to happen. Probably just the brine alone would make you ill in that case."

How the Lake and Its Life Grows

Water flows into Lake Vida when it's warm enough for nearby glaciers to melt a little, and this isn't guaranteed each year. When it's warm, some of the ice on top of the lake also melts. Doran says about 3 feet (1 meter) of water was on top of the lake's ice in the 2001–2002 summer, but none during the colder 2002–2003 summer. Water brings in sediment, which means ice cores from the lake have layers of it from warm years.

During the warm summers, a small ecosystem starts in the water on top of the old ice. These creatures, of course, are frozen into ice that forms in the fall. Even if the next summer isn't warm enough for the ice to completely melt, the sun heats the ice and sediment enough 3 to 6 feet (1 to 2 meters) down to get the ecosystem going again, using the liquid water in the ice. Below about 6 feet (2 meters), the ice never melts, and the microbes there are in the deep freezer, but apparently ready to come back to life if the ice ever melts. The radar showed water at the very bottom, which has to be extremely salty not to freeze.

Vida Drilling Will Use New Space Technology

Scientists plan to drill into Lake Vida to see what's in the extremely salty water at the very bottom. They just might learn that life can stand more salt than we now believe. The big concern is to ensure that the drilling equipment doesn't contaminate the ice and water below it with anything. Developing and using the needed sterile system will be part of the new NASA Astrobiology Science and Technology for Exploring Planets program.

With this program, NASA is working with the National Oceanic and Atmospheric Administration, and the National Science foundation to develop the technology and scientific understanding needed to search for life on other planets.

Lakes Are Discovered Under Antarctica's Ice

As far back as the middle of the 1970s, a few air-borne studies with radar that looks into the ice led scientists to think that unfrozen water lay under Antarctica's ice in a few places, especially under the Russian Vostok Station on the Polar Plateau. In 1996, English and Russian scientists discovered that as many as 70 lakes are under the ice, with "Lake Vostok" by far the largest at about 3,860 square miles (10,000 square kilometers). The lake is usually described as being about the same size as Lake Ontario.

This started some people wondering whether the lakes might hold life. Before anyone realized that a huge lake lay under their station, the Russians had started drilling an ice core at Vostok to study past climates. With the speculation that something might be living in the lake, the Russians stopped drilling at a depth estimated to be about 300 feet (100 meters) above the lake. The Russians sent samples from the parts of the core from near the bottom, but not from the very bottom, to scientists in other nations, including the United States, to study.

Scientists Find Life in Vostok Core

In December 1999, two research teams, one led by David M. Karl of the University of Hawaii and John C. Priscu of Montana State University, reported they had found bacteria in ice from about 11,700 feet (3,600 meters) below the surface.

"From a biologist's perspective, this is the Holy Grail of lake biology," Priscu said. "Our findings indicate that the microbial world has few limits on our planet. You don't have to leave the planet to study this completely unexplored system, but the samples sure aren't easy to get."

Ice Chips

The bacteria found in the Vostok core in 1999 are ones commonly associated with soils, and could have been blown on bits of soil from South America onto the ice. If so, the microbes could be more than half a million years old. Another possibility is they originated in the lake and became trapped as lake water refroze to the bottom of the overlying glacier.

How to Study Lake's Life Is the Question

Once scientists were sure that Lake Vostok holds life everyone involved agreed that the last thing they wanted to do is contaminate the lake. On the other hand, NASA sees Vostok as the perfect place to test devices that could be loaded onto a spacecraft and sent off to explore under the ice of Europa.

Scientists Learn How Ice, Lake Interact

During the 2000–2001 research season, the National Science Foundation picked up the tab for detailed radar mapping and other studies of the ice over and around Lake Vostok. Using the data, Robin E. Bell of Columbia University's Lamont-Doherty Earth

Observatory and her colleagues, discovered how the lake exchanges water with the moving ice above it. This means that while the lake has been sitting there for perhaps millions of years, all the water in it ends up being replaced every 13,000 years.

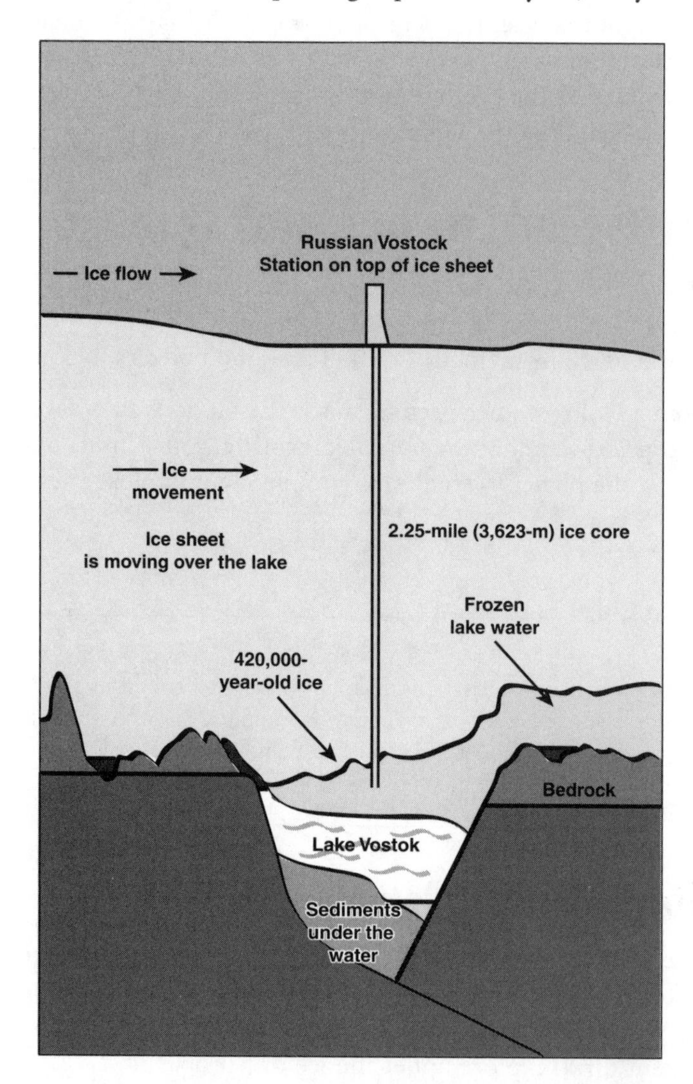

Cross-section view of ice and Lake Vostok.

(Information from Ravi Rajakumar, Columbia University Earth Institute)

If you could go down to Lake Vostok, you almost surely would not be able to row around on the lake under a dome of ice. The bottom of the ice sheet is touching the top of Lake Vostok's water.

Bell explained that the ice sheet is moving from west to east across the lake, "kind of like a conveyor belt" at 6 to 9 feet (2 to 3 meters) a year. As the ice moves across the lake, some of the water on top of the lake freezes to it and is carried to the east. Water that freezes to the bottom of the ice sheet is replenished, either by ice on the bottom of the ice sheet melting, or water from elsewhere flowing under the ice into the lake.

This means that researchers can learn much about what's in the lake by drilling down to ice that has moved to the east with lake water, Bell said. "These frozen lake water samples will record the passage of the ice sheet and the processes across the lake."

Ice Chips

The Russian Vostok Station is 11,484 feet (500 meters) above sea level, which means anyone who goes there faces the danger of altitude sickness. The high and low temperatures in January, the warmest months, are –20°F (–29°C) and –32°F (–35°C). The United States has built a small, separate camp near the Russian station.

Anything in Lake at Least 400,000 Years Old

Even though she estimates that all of the water in the lake is replaced every 13,000 years, Bell said, "Lake Vostok is absolutely devoid of interference. The youngest water in it is 400,000 years old. It doesn't know anything of human beings, fossil fuels, or plastics. It is a window into life forms and climates of primordial eras."

How can this be? Any water in the lake comes from the bottom of the ice sheet. Scientists calculate that a snowflake, a microbe blown in by the wind, or a cigarette butt dropped by one of the Russian scientists at Vostok needs 400,000 years to work to the bottom of the ice sheet. No, things don't worm their way through the ice. The ice sheet is creeping toward the coast as snow continues falling on the top.

New Round of Ice Studies Begins

Ice from the bottom 35 feet (12 meters) of the core were in a storage trench at Vostok until the 2001–2002 season when it was flown to the U.S. McMurdo Station and then put on a ship to France. Pieces of the ice are being parceled out to French, Russian, and American scientists.

Those who have seen the ice have commented that it's much clearer than ice from higher up in the core. This probably means it's made of water from the lake and could contain different life than the ice examined in 1999. The bacteria in the ice studied in 1999 could have been on their way down to the bottom, not from the lake itself.

Lessons from the Ice

We'll have to see whether drilling into Lake Vida to discover what's in the salty water at the bottom, or sampling the water of Lake Vostok will tell us anything about possible life elsewhere in the universe. But "the discoveries of life in Antarctica's ice are only a small part of a much larger and more globally important, multifaceted story," Priscu said. "The earth's biosphere is larger than we had ever imagined, and the microbial world has few limits on our planet and, possibly, others."

The Least You Need to Know

- Scientists think understanding life in Antarctica could aid the search for life on other planets.
- Microbes found frozen in Lake Vida show life can survive at lower temperatures than expected.
- Scientists have known since the 1970s that lakes are under more than 2 miles (3 kilometers) of Antarctic ice.
- Lake Vostok, about the size of Lake Ontario, is the largest of these lakes.
- Scientists are looking for ways to learn what, if anything, lives in Lake Vostok without contaminating it.

Part 6

The Polar Regions and the Rest of the World

The polar regions are the earth's air conditioners. To have a better idea of what might happen as the earth's climate warms, scientists need to have a good idea of how these air conditioners work. There can be little doubt that the earth's climate is changing, and some of the biggest changes are being seen around the Arctic. Scientists, as well as the people who live in the Arctic, want to know what's likely to happen if the warming continues. Scientists also want to learn how changes in the Arctic and Antarctic will affect the rest of the world; how the air conditioners will work in a warmer world.

Chapter 25

Polar Connections with the Rest of the World

In This Chapter

- An Antarctic expedition runs into El Niño
- The oceans and atmosphere carry tropical heat to the polar regions
- The Antarctic cools the earth more than the Arctic
- Antarctic upper air winds set the stage for the ozone hole
- Oscillations are a key part of the weather picture.

Without the Arctic and Antarctic, the earth would be warmer than it is. In this chapter, we'll look at how the polar regions work as earth's air conditioners, and also at a few of the connections between the polar regions and the rest of the earth and how scientists are working to learn more about them.

El Niño Reaches Antarctica

When the 13 men and women of the U.S. International Trans-Antarctic Science Expedition (U.S. ITASE) set out from the Byrd Surface Camp for the South Pole in late November 2002, the things they were looking for included evidence of how the global weather pattern known as *El Niño* affects Antarctica. They ran into more El Niño than they wanted.

The expedition ran into unexpected deep snow at the beginning and near the end of the 800-mile trip. The team's two tractors pulled sled trains with the huts the members lived in, scientific equipment, food, and fuel. At the beginning the team managed to go only 25 miles in two days before getting stuck in deep, soft snow. After turning back and fitting one of the tractors with wider treads and the sled with the heaviest load with better runners, they set off again. Near the end of their journey, they had to double up with the two tractors to pull half the train out of soft snow, and go back for the other half.

Polar Talk

Ocean-atmospheric interactions in the tropical Pacific that are characterized by episodes of unusually high or low sea surface temperatures in the tropical eastern Pacific and associated with large-scale swings in surface air pressure between the western and central tropical Pacific are called the El Niño-Southern Oscillation (ENSO). The phase with warm water in the eastern Pacific is known as **El Niño** and the phase with cool water there is known as *La Niña*.

Paul Mayewski, the expedition's leader, knew that El Niño, which affects weather far away from the tropical Pacific, sometimes brings more snow than normal to the part of Antarctica the expedition traversed, but he never expected such a large effect. Data the expedition collected could help clear up questions about how the ocean and atmosphere of the Tropical Pacific helps determine how much snow falls in Antarctica.

Global Connections Are Often Subtle

Connections between the polar regions and other parts of the world are most obvious to those who live in the northern parts of Northern Hemisphere continents. When winds from the north send temperatures plunging to zero in Minnesota or Moscow, it's easy to believe that the cold is coming straight from the North Pole.

Only in recent years have meteorologists made the connection between an eastward shift of the Pacific's warmest water with lower January heating bills in Wyoming and a wet winter in Louisiana. These are two of the well-established facets of the El Niño phase of the ENSO pattern.

In 2001, Kevin Trenberth and Julie Caron of the National Center for Atmospheric Research analyzed huge amounts of data and concluded that the atmosphere handles 78 percent of the heat transported from the tropics to the Arctic and Antarctic. In their *Journal of Climate* report, Trenberth and Caron say that at 35 degrees latitude, where the greatest amounts of heat are headed toward the poles, the atmosphere is carrying 78 percent in the Northern Hemisphere, and 92 percent in the Southern Hemisphere.

The global currents we looked at in Chapter 5 do the ocean's share of this heat hauling, and winds do some of the atmosphere's share, but not all. Storms make up the difference. The large storms that move across the middle latitudes in both hemispheres pump warm air toward the poles and cold air toward the equator. Tropical cyclones such as hurricanes and typhoons carry heat poleward.

Ice Chips

Kevin Trenberth and Julie Caron estimate that the oceans and atmosphere carry amounts of heat to the polar regions each year equal to the energy produced by 5 million power stations, each generating 1,000 megawatts. This is as big as power stations get.

An Extreme Continent Has Extreme Effects

An icy continent occupies almost all the Antarctic (the area south of the Antarctic Circle) while most of the Arctic is ocean. This ensures that the two regions go about their work of being the earth's air conditioners in different ways. Antarctica has more ice to reflect away sunlight during the summer than the Arctic, where except for Greenland and few small ice caps on islands, the snow melts as does a great deal of the sea ice.

Explorations

Even though the Antarctic is colder than the Arctic, more people feel the Arctic's chill because the Southern Ocean warms Antarctica's frigid air. Two cities tell the tale. Punta Arenas, Chile, the closest city to Antarctica, has an average July (mid-winter) low of 31°F (−1°C). Edmonton, Alberta, Canada, about the same distance from the North Pole as Punta Arenas from the South, has an average January (mid-winter) low of 1°F (−17°C). Arctic air warms little traveling over land to Edmonton.

If you stand at the North Pole, water warmer than 32°F (0°C) is around 800 feet (250 meters) under your feet, thanks to the warm Atlantic Ocean flowing around the Arctic Ocean (see Chapter 5). If you stand at the South Pole you're standing on around 9,000 feet (2,750 meters) of ice that's much colder than 32°F (0°C). The colder Antarctic is earth's biggest air conditioner.

To look at its weather, think of Antarctica as a huge block of ice, with mountains sticking out of the ice in a few places to channel winds. In Chapter 2, we saw how when frigid air builds up over Antarctica, the heavy air begins flowing downhill as katabatic winds, which blast out over the oceans around the continent.

Ice Chips

An example of Antarctica's global effects: When David H. Bromwich and his colleagues from the Byrd Polar Research Center at Ohio State University ran a sophisticated computer model that simulates global climate, they found that if they replaced the Southern Ocean's sea ice with open water, it triggered a chain of atmospheric changes that spread like a wave, including effects such as a delay in the start of China's winter monsoon winds.

As winds blow air away from Antarctica, more air sinks from above to replace it. The sinking air warms, but someone standing on the ice doesn't feel warm air because the extremely frigid ice chills the air next to it. This creates an inversion, warm air atop cold air. At the South Pole in the middle of winter, air about 1,000 feet (300 meters) up averages about 50°F (27°C) warmer than the air around chest high of someone who ventures out into the –85°F (–65°C) cold.

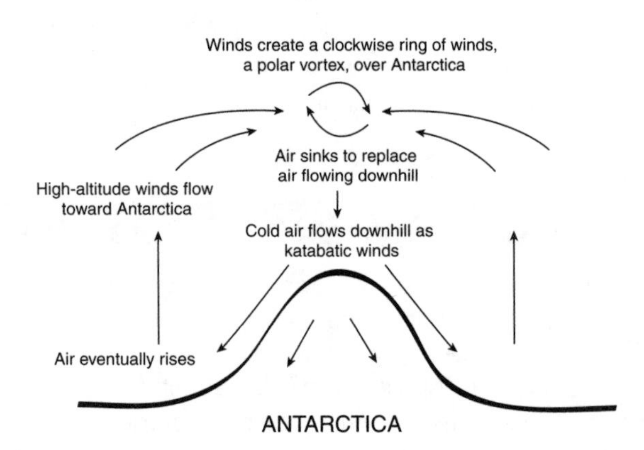

A simplified view of Antarctic air flow.

Because the air that's flowing in high above the ice is on a rotating Earth, it doesn't follow a straight line; instead, it makes a clockwise circle around Antarctica. It creates what's known as the polar vortex—that is, a ring of wind circling high above Antarctica. It's not a perfect circle, and sometimes it takes different shapes with one side or the other pushed in or bulging out, but it continues until the sun rises, warms the air, and slows the flow of high-altitude air toward Antarctica.

How the Ozone Hole Came About

Scientists have known since 1881 that stratospheric *ozone* absorbs much of the sun's most damaging ultraviolet light, but until the 1960s, no worried about ozone. In the 1970s, scientists discovered that various substances could harm ozone in the stratosphere, especially man-made chemicals called *CFCs*, which were rising into the stratosphere. This discovery prompted some actions to curb CFCs, such as the 1979 ban of spray cans using CFCs by the United States, Canada, Norway, and Sweden. Companies that made CFCs and similar chemicals began working on substitutes, but they seemed to have plenty of time. Scientists estimated that CFCs would destroy only 5 percent of the stratosphere's ozone by 2050.

Polar Talk

Ozone is an almost colorless, gaseous form of oxygen consisting of three oxygen atoms. Molecular oxygen, which makes up about 21 percent of Earth's atmosphere, has two oxygen atoms. **CFC** is an abbreviation for chlorofluorocarbons, which are synthetic chemical compounds that contain chlorine, fluorine, and carbon. They were invented to be refrigerants, but came to have many other uses, including as solvents for electronic parts. They are nonpoisonous and nonflammable, and replaced other, dangerous substances in refrigerators. Under normal conditions near the ground, they do not break into their components, including chlorine.

Discovery of Ozone Hole Shocks Scientists

A British Antarctic Survey announcement in 1985 that its Halley Station on the Antarctic Coast had been observing a huge reduction in the amount of ozone overhead in the spring shocked scientists. In October, the British said, the air above Antarctica had about 35 percent less ozone than the average for the 1960s. The figures didn't seem to make sense, but other measurements confirmed the results.

To find out what was happening, researchers went to Antarctica in 1986 and 1987 and begin making measurements. Those measurements, including data collected by NASA ER-2 airplanes that flew into the Antarctic stratosphere, led the scientists to conclude that chlorine and bromine from man-made substances trapped in Antarctica's wintertime polar vortex were causing the ozone hole.

Their findings were so convincing that they led the two dozen nations that manufactured or used more than 80 percent of the world's ozone-destroying substances to agree to begin phasing them out, especially CFCs. Later agreements ended all CFC production in 1996, but since CFCs stay in the atmosphere for decades, the Antarctic ozone hole isn't expected to disappear until at least the middle of the twenty-first century.

How the Ozone Hole Works

The polar vortex that forms over Antarctica in the winter cuts off the flow of air from the north, allowing clouds of nitric acid and water to form inside the "container" the vortex creates. These "polar stratospheric clouds" provide a place for the substances that destroy ozone to begin the first stage of a two-step chemical process. The first stage goes on during winter's polar darkness. It sets the stage, but doesn't destroy ozone.

Sunlight triggers the second step, which is why the ozone hole doesn't appear until the sun begins to rise over Antarctica. As ozone destruction begins, the vortex is holding everything together. It's almost like a pot that ozone and its enemies have both been thrown into. Ozone destruction continues until the air warms enough for the vortex to begin breaking down. When this happens, the ozone-destroying chemicals are spread around and air with more ozone in it begins to flow over Antarctica. Ozone levels return to normal, usually by early January.

Ice Chips

Ice cores in both Greenland and Antarctica contain radioactivity from open-air nuclear bomb testing from the mid-1950s until the mid-1960s when atmospheric testing ended. Less than two years after the Russian nuclear power station at Chernobyl exploded, its radioactivity showed up at the South Pole.

The Arctic Also Imports Air Pollution

An ozone hole doesn't form over the Arctic regularly because in most years a strong stratospheric vortex doesn't form there because the Arctic isn't as cold, and the Northern Hemisphere's large mountains disrupt the flow of air around the Arctic.

In the mid-1990s, scientists were surprised to find higher levels than expected of nitrogen oxides (called NOx) in the air above the snow at the National Science Foundation's Summit research camp on Greenland's ice sheet. Greenland's air is as free of pollution as any in the Northern Hemisphere; NOx is the kind of stuff you expect to find in the air over Los Angeles or other big cities.

Roger Bales of the University of Arizona, one of the scientists studying the air at Summit, says rain or snow wash NOx from the air as nitric acid within a few days of its coming from smokestacks or tailpipes. Some of the NOx put into the air over North America ends up as nitric acid in Greenland's snow. When sunlight shines on that snow, some of this nitric acid is changed back to NOx and released to the atmosphere.

"The snow is a chemical reactor, just like the air is. It's a concentrated reactor at the surface of ice grains," Bales says. What goes on in Greenland's ice is important because it gives pollution a "second chance" to affect the atmosphere—another way that the atmosphere links the world to the polar regions.

Oscillations Seen to Make the Climate Go Around

We've talked of the polar regions as the earth's air conditioners, but they do more than take in air and ocean water, cool them, and send them back toward the equator. In the last decades of the twentieth century, scientists recognized that Earth has *Arctic and Antarctic oscillations;* changing patterns that create week-to-week, year-to-year, and decade-to-decade weather swings.

Arctic Oscillation Felt Around Northern Hemisphere

Americans are used to hearing how El Niño is likely to bring storms to the West Coast, or winter rain to the states across the southern edge of the nation. But for the northern part of the United States, the Arctic Oscillation is more important than El Niño. There's one important difference. The El Niño-Southern Oscillation operates on a three to seven or so year schedule, taking months, usually more than a year, to swing from one phase to another. The Arctic Oscillation can swing back and forth in a week. But, and this is important, it can go years, maybe decades, favoring one of its phases. That is, it will make weekly swings, but on the average, one phase will dominate.

John Wallace of the University of Washington notes that while it's called the "Arctic Oscillation," the oscillation's causes as well as effects extend well beyond the Arctic to the entire Northern Hemisphere, but it is centered on the Arctic. From the late 1970s into the 1990s, the Arctic Oscillation has favored its positive phase, which could account for a great deal of the warming during the winters in Europe and central Canada.

Polar Talk

Antarctic and Arctic oscillations are weather patterns of air pressures and winds centered over the Antarctic or Arctic with two phases. The positive phase is characterized by lower air pressure and strong winds ringing the polar region. The negative phase is characterized by higher air pressure and weaker winds around the polar region.

Ice Chips

What causes the Arctic Oscillation? Richard Kerr, writing in *Science,* put it this way: "It's a natural mode of the atmosphere, just as a drum has a natural mode of vibration. Hit a drum almost anywhere with almost anything, and much the same sound comes out; hit the atmosphere—with random jostlings, sunlight-reflecting volcanic ash in the stratosphere, variations in solar brightness, or added greenhouse gases—and it will oscillate with the pattern of the AO."

The Arctic Oscillation could also account for the reduction of sea ice in the middle of the Arctic Ocean from the 1970s into the early 2000s, Wallace says. Normally an area of high air pressure, with winds going clockwise around it, sits roughly over the middle of the Arctic Ocean. These winds push the sea ice in a rough circle known as the "Beaufort Gyre." With the Oscillation favoring the positive phase, these winds have weakened. Instead of going in a circle, more Arctic Ocean ice drifts to the east and into the Atlantic between Greenland and Europe.

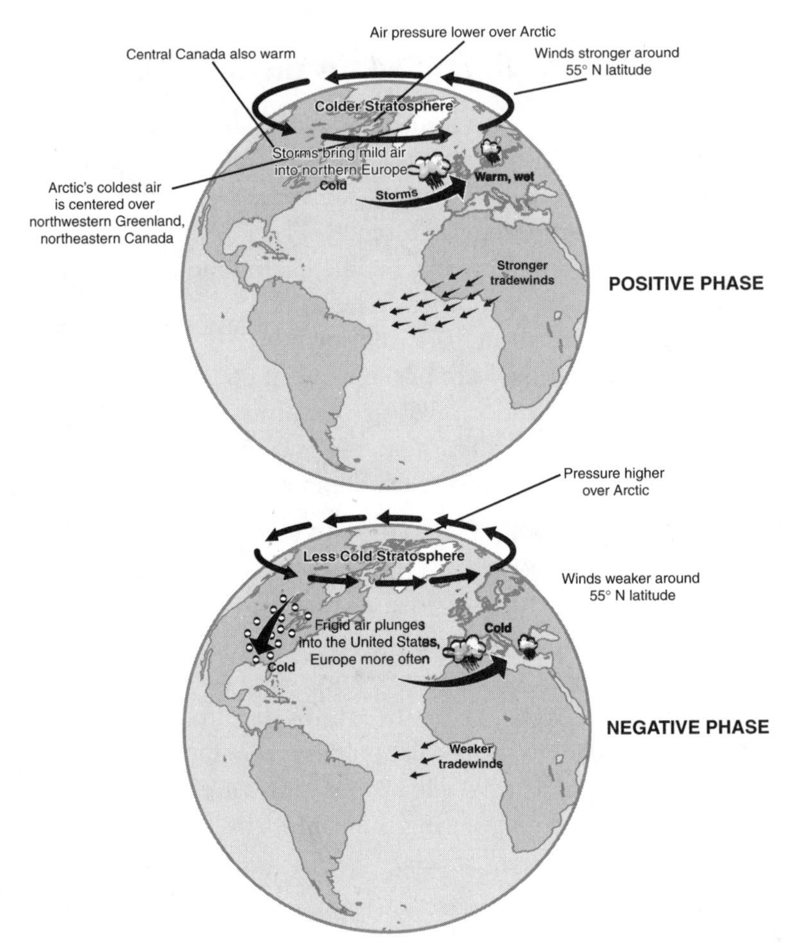

The two phases of the Arctic Oscillation.

Oscillations Can Help Solve Climate Puzzles

In Chapter 28, we'll look at the idea that global climate change, caused at least in part by gases humans are adding to the atmosphere, could be one of the factors that's banging on the Arctic Oscillation's "drum" today.

Wallace says the Arctic and Antarctic oscillations are not "just ad hoc patterns," but exist for good reasons that scientists are beginning to understand. Studies of the oscillations are giving scientists new insights into how the earth's climate works.

We see then that the growing knowledge of the Arctic and Antarctic oscillations is important because they account for some, maybe many, of the changes we have been seeing in the polar regions in recent years. As scientists try to understand how human actions are changing the world's climate, one of their tasks is to separate changes being caused by the various natural oscillations and the changes humans are causing. Even more important will be discovering how the natural oscillations are responding to climate changes and how they are likely to respond in the future.

The Least You Need to Know

- Effects of the pattern of ocean-atmosphere shifts known as El Niño are felt in Antarctica.
- The atmosphere carries about 78 percent of the heat that moves from the tropics to the polar regions; the oceans carry the rest.
- Storms are one of the main ways the atmosphere moves warm air to the polar regions and cold air away from them.
- The polar vortex of winds and substances made by humans are the main causes of Antarctica's ozone hole.
- International bans on ozone-destroying substances could close the ozone hole after the middle of the twenty-first century.
- Learning more about the Arctic and Antarctic Oscillations should improve understanding of global climate.

Chapter 26

Ice Tells Climate Stories

In This Chapter

- Scientists learn to read climate history in ice
- How scientists figure out the dates when ice formed
- Ice cores can be a sensitive, historical thermometer
- Greenland ice cores show that climate isn't as steady as believed
- Cores drilled during an Antarctic crossing promise to solve climate mysteries

Scientists can confidently make statements such as "Earth was in an ice age 35,000 years ago" because they have learned how to use ordinary things such as fossil pollen, or mud that settled on the bottom of a lake or ocean thousands of years ago to tell what the climate was like when the fossils were living creatures or the mud formed. One of the best records of past climates, however, turns out to be ice that's been around for a long time deep within the Greenland and Antarctic ice sheets, or in glaciers around the world. In this chapter, we look at how ice stores climate history and how scientists read this story.

Glaciers and Ice Sheets Store Climate History

In Chapter 4, we saw how snow that piles up slowly becomes glacier ice. Because glaciers move so slowly toward where the ice melts or falls into the sea, the ice deep within a glacier or ice cap could be hundreds or even thousands of years old.

After World War II, scientists began to wonder what they could learn if they could examine the ice deep inside a glacier or an ice sheet. If nothing else, the bubbles of air and dust that had been trapped in the ice centuries ago might offer clues to climate. In the early 1960s, a Danish geochemist, Willi Dansgaard, worked out a way to use the very atoms of oxygen in the ice's water molecules to measure the temperatures when the snow fell. This made ice an even more tempting source of climate history.

Ice Chips

Camp Century was an experimental U.S. Army base built in tunnels in Greenland's Ice Sheet about 100 miles east of the U.S. Thule Air Force Base. It opened in 1959 and closed in 1966. Its small nuclear power plant supplied enough energy for as many as 200 men to live almost as they would at a base in the United States, with roomy dormitories, hot showers, and even enlisted men's and officers' clubs.

Polar Talk

An **ice core** is a cylindrical piece of ice removed from an ice sheet, ice cap, or glacier by a drill designed to do this. The term can refer to all of the ice removed from a hole drilled to the bottom of an ice sheet, or to individual 3-foot (1-meter) pieces.

The only way to get undisturbed ice from deep within an ice sheet is to drill a hole and pull out some ice. In 1966, the U.S. Army's Cold Regions Research and Engineering Laboratory drilled down to bedrock at the Army's Camp Century in Greenland, retrieving an *ice core*. Dansgaard and his group used his new technique to study the ice, discovering, among other things, that past climates hadn't been as steady as many believed. Scientists disputed this finding, but not the proof that ice can be pulled from deep within ice sheets and that it has stories to tell.

Researchers Pull History Out of the Ice

An ice-coring drill is a piece of open pipe with teeth around the edge at one end. Such a pipe, called a drill barrel, that's maybe 20 feet (6 meters) long, and 5 or 6 inches (12 to 15 centimeters) in diameter, is lowered into a hole in the ice. An electric motor turns the pipe, teeth cut into the ice, and chips go up spirals on the outside of the barrel as a 5-inch-wide cylinder of ice slides up inside. When the ice cylinder is 10 or 20 feet long, the drill stops, "fingers" snap up to hold the bottom of the ice cylinder in the barrel, and the barrel is pulled up, where the drilling crew removes the core, lowers the drill, and does it over and over.

When ice cores are pulled up, you hear tiny "pops" as bubbles of high-pressure air locked in the ice burst. In fact, sometimes cores from deep in the ice explode—giving researchers an icy jigsaw puzzle. The ice's pressure is also trying to close the drill hole. To prevent this, drillers fill the hole with a liquid that's the same density as water, but doesn't freeze, such as butyl acetate.

Someone has to carefully log information about each piece of the core and label it. To learn anything from ice cores, scientists can't have any doubt about where a particular core fits into the sequence. Even worse would be to allow ice cores to melt. It's common to keep cores in a –22°F (–30°C) trench for a year or more to allow the ice to "relax" as high-pressure bubbles expand until the air pressure inside matches the outside air pressure.

Reading Ice Cores Is a Complicated Science

One of the first things you notice when you look at a piece of ice core from Greenland or Antarctica with light shining through it, is that it has light and dark bands. In both places, snow falls all year, and the sun doesn't set in the summer. This helps make the snow different in summer and winter.

Although the midday sun doesn't melt snow, it heats it enough to turn some ice crystals directly into water vapor that goes into the air. When the midnight sun dips low in the sky, water vapor in the air condenses into ice crystals, called *hoar frost* on the snow. The makes summer snow less dense than winter snow. The summer snow shows up as light bands while the winter snow is dark. Scientists use these layers like *tree rings* to count the years.

Polar Talk

Hoar frost is a deposit of interlocking ice crystals formed by direct deposition of water vapor on objects, including snow or grass, freely exposed to the air. A **tree ring** is a sheath of cells forming a circle around a cross section of a tree limb or trunk. The cells are formed by each spring's growth spurt. Thus, they can be counted to determine the age of a tree.

The summer-winter bands sometimes can be seen in ice more than 1,000 years old. Even then, differences remain that other techniques can detect. But as you go farther back in time, the ice's pressure squashes years together; dates are less precise.

Volcanic eruptions can give precise dates to parts of ice cores. A large volcano, even on the other side of the world, spews large sulfur and other particles into the air that fall on the ice. The sulfur becomes sulfuric acid in the air, and this is seen on the ice along with microscopic particles from the volcano.

Explorations

Lead levels in Greenland ice cores tell a story of economic ups and downs. Before the 1930s, ore smelting and coal burning accounted for most of the lead in the atmosphere. Lead amounts in ice formed during the Great Depression of the 1930s dropped, but rose sharply during the boom of the late 1940s and early 1950s, when leaded gasoline became popular. Stricter U.S. pollution controls in the 1970s brought down lead levels. Lead pollution isn't only recent. Greenland ice cores show traces of lead from around 2,000 years ago, probably from smelting by the Romans.

Greater amounts of dust fall with the snow during widespread or prolonged dry periods. Often, the chemical composition of dust in a core tells researchers where it came from, enabling them to map ancient droughts. Some chemicals found in cores tell scientists about the abundance of certain kinds of ocean life at the time. *Isotopes* of various elements, such as Beryllium10, which are created by cosmic rays, track solar cycles.

Polar Talk

Isotopes are forms of atoms of an element that differ in atomic weight. All atoms consist of a nucleus made of protons, with a positive charge; and neutrons, with no charge; surrounded by a cloud of electrons, with negative charge. The number of protons and electrons are the same, but the number of neutrons can differ in atoms of an element. Oxygen, for example, has eight protons, but can have eight, nine, or ten neutrons, giving it three isotopes.

Ice Cores Offer Records of Past Temperatures

Water molecules are made of an oxygen atom and two hydrogen atoms. Even though all water molecules are made of these two elements, all water molecules aren't alike because both oxygen and hydrogen have more than one isotope. Around 99.8 percent of oxygen atoms are oxygen-16 with a nucleus made of 8 protons and 8 neutrons. Next most common is oxygen-18 with a nucleus of 8 protons, but 10 neutrons. It has an *atomic mass number* of 18.

Polar Talk

Atomic mass number is the sum of the protons and neutrons in an atom.

Hydrogen also has isotopes, with by far the most common being hydrogen-1, with only one proton. Hydrogen-2 has one proton and one neutron.

All Water Doesn't Weigh the Same

Almost all water molecules are one atom of oxygen-16 and 2 atoms of hydrogen-1 for a total atomic mass of 18. Other combinations of oxygen and hydrogen atoms create heavy water. Ordinary and heavy water act the same chemically, but the tiny additional weight of heavy water affects how it evaporates and condenses. Heavy water is a little slower to evaporate from the oceans, but some becomes water vapor. When condensation begins, heavy water is the first to join the water droplets or ice crystals and fall. As water vapor in the air becomes colder, it has less heavy water.

Willi Dansgaard's great breakthrough was discovering that the number of oxygen-18 atoms in snow or ice compared with the number of oxygen-16 atoms is a good indication of how cold the air was when the snow fell. His basic finding was that the fewer oxygen-18 atoms, the colder the air. In a paper, published in the European scientific journal *Tellus* in 1964, he showed that isotope ratios correlated very well with average temperatures.

Ice Chips

Dansgaard's finding that oxygen isotope ratios are a measure of temperatures when snow fell went back to Harold Urey's discovery in the 1950s that similar ratios could be used to derive the ocean temperature at the time ocean fossils were living creatures, taking up oxygen from the water. His University of Chicago group used the method to find plausible ocean temperatures back more than 100 million years ago.

Americans and Europeans Drill Deep into Greenland

Between 1989 and 1993, the United States and a consortium of European nations drilled two ice cores only 18 miles (29 kilometers) apart at the summit of the Greenland

Ice Sheet. The ice here is about 10,000 feet (3,050 meters) deep, and holds a climate record going back more than 100,000 years into the warm period before the last *ice age* began. Drilling cores that close together would ensure that the results were typical for that part of the ice sheet, if they agreed. The American effort at the Summit Camp was the Greenland Ice Sheet Project 2 (GISP2) while the European effort was the GReenland Ice core Project (GRIP), drilled at the GRIP Camp.

Polar Talk

An **ice age** is a period in the earth's history when ice sheets cover vast regions of land. An ice age typically lasts about 100,000 years.

Cores from Greenland Agree and Disagree

The cores agreed remarkably well back to 113,000 years ago. But diverged sharply on what happened from 113,000 years ago to around 125,000 years ago. The European core shows the climate making sharp swings in temperatures and precipitation during this period, but the U.S. core shows little change. Scientists are sure the two disagree because ice at the bottom of one of the cores, or maybe both, had been deformed as the ice slid across the rock under it.

What was going on from 113,000 to 125,000 years ago is important because the earth was then warm, in fact a little warmer than now, but it was moving into the ice ages that ended about 10,000 years ago. That is, the climate was changing. Even though that change was a cooling, what happened then could give some clues to what could happen when the change is going in the other direction. To discover what happened more than 113,000 years ago, the Europeans built a North GRIP camp 228 miles north of the Summit and GRIP drilling sites, and began drilling in 1996. After many problems, the Europeans hoped to finish in 2003.

Cores Show Climate Change Isn't Steady

Until the 1980s, most climate scientists thought that changes from ice ages to *interglacial* periods occurred slowly and probably smoothly. Human experience seemed to show that while the weather swings from floods to drought, or hot to cold, from year to year, the average weather, which is what we mean by climate, holds a steady course. The first results from the 1980s Greenland ice cores cast doubt on the idea of steady climate change, but the evidence wasn't yet enough to overthrow the old view.

Polar Talk

An **interglacial** is a period in Earth's history when ice sheets generally disappear. An interglacial period typically lasts 10,000 to 20,000 years. Earth has been in an interglacial for about 10,000 years.

Even with the remaining questions, the GISP2 and GRIP cores clearly show abrupt climate changes. Andrew Mayewski, the chief scientist for the GISP2 program, writes in his book, *The Ice Chronicles: The Quest to Understand Global Climate Change.* "The well-dated record that resulted from the GISP2 and GRIP ice cores changed forever and quite dramatically the view held earlier that natural climate variability operates on a slow time scale."

What kind of events are we talking about? Ice cores and other evidence "increasingly suggest that about 11,500 years ago, during a period called the Younger Dryas, global temperatures fell by up to 16° within a decade and rainfall halved. Things stayed that way for more that 1,000 years," Richard Alley of Pennsylvania State University, told the American Geophysical Union annual meeting in December 2001. What would that mean? To take one example, a 16 degree drop would mean winters in Wilmington, North Carolina, becoming more like those in Boston, Massachusetts.

Climate Switches Could Be in Polar Regions

Alley says that until ice cores showed otherwise, scientists had thought of climate changing much as though a dimmer switch, which was always turned slowly, controlled it. The ice cores show that climate also has "on-off" switches. At times, Alley says, the records make it look like a mischievous child was playing with the switch. Scientists have reasons to think that some of these on-off switches could be in the polar regions.

Europeans Drilling New Antarctic Cores

The new European Project for Ice Coring in Antarctica (EPICA) hopes to answer questions about climate switches, among other things, with two cores it is drilling.

In December 2002, an EPICA core at the Dome Concordia site on the Polar Plateau reached 9,842 feet (3,000 meters) where the ice could be 700,000 years old. The other core is being drilled at Kohnen, which is south of the Atlantic Ocean. In addition to looking for Antarctic climate switches, the Europeans are trying to see how climate change in the Northern and Southern Hemispheres are linked, and whether the last 10,000 years has been unusually stable compared with past climates.

New Generation Coring Could Answer Big Questions

In Chapter 25, we saw how the 2002–2003 U.S. International Trans-Antarctic Scientific Expedition (U.S. ITASE) lead by Andrew Mayewski, ran into deep snow, which Mayewski is sure was one example of the effects of El Niño on Antarctica. As it turns out, the eight ice cores the expedition extracted on its 800-mile (1,287-kilometer) trip should help show just how El Niño affects the part of Antarctica they crossed. The cores were deep enough to record the area's climate going back about 200 years.

Cores Fill In for Unavailable Weather Data

Scientists who want to know the effects of El Niño in places such as southern California use the detailed weather records to see what happened during El Niño years. This doesn't work for Antarctica, which has few weather records.

The expedition's ice cores, however, should show season-by-season weather going back 200 years. This data could show whether there are consistent patterns of temperatures and snow during El Niño years. If there does turn out to be a consistent pattern, Antarctica could be the way to trace the history of El Niño far back in time, Mayewski says. Right now, *paleoclimatologists* don't have much to go on as they try to figure out whether El Niño has changed over the centuries. Antarctic ice cores, however, could show any El Niño changes.

> **Polar Talk**
>
> A **paleoclimatologist** is a scientist who studies climates of the past for which written weather records are not available, using data from ice cores or other sources of data on past climates.

Mayewski says data-gathering methods, including ice coring along traverse routes, U.S. ITASE expeditions are using promise "to light up the black hole of Antarctic climate data. We hope to change Antarctica from the most poorly understood part of the world to the very best understood."

The Least You Need to Know

- Glaciers and ice sheets are storehouses of climate data.
- Special "coring" drills are used to pull cylinders of ice from deep within ice sheets.
- Scientists can use various methods to learn the age of ice in cores.
- The amounts of different forms of oxygen atoms in ice tell researchers the temperature when the snow that made the ice fell.
- Ice cores show that Earth's climate has made sudden shifts in the distant past.
- New ice cores being drilled in Antarctica could answer many climate questions.

Chapter 27

Is the Ice Melting?

In This Chapter

- A quick look at the different kinds of polar ice
- Why melting polar ice is important
- What's going on in Antarctica as huge icebergs break off and ice shelves collapse
- What's happening to the Greenland Ice Sheet
- In some areas the amounts of sea ice is increasing while it's decreasing in others

Polar ice is getting into the news regularly as we hear of icebergs the size of Long Island breaking off from Antarctica, and sea ice disappearing from the Arctic Ocean. In this chapter, we'll look at how much scientists know about what's happening to the different kinds of polar ice at both ends of the earth, and what they are trying to learn.

Why We Care About Polar Ice

If the water now locked up as ice in Antarctica and Greenland melted, sea levels would rise drastically—by 215 feet (65 meters) if all of Antarctica melted, for instance. Fortunately even the most extreme warming scenarios don't warm Antarctica enough for this to happen for centuries. If Greenland's ice sheet melted, sea levels would rise by about 21 feet (6.5 meters), but this, too, is not considered likely in the lifetime of anyone living now. Although the chances of a complete melting of ice in the next few centuries seems extremely remote, this doesn't mean we have nothing to be concerned about.

More Than Melting Is Involved in Ice Loss

Glaciers, ice caps, and ice sheets grow, move, and shrink. The real question we should be asking is not "Is the ice melting?" Instead we should be trying to discover how a glacier's or ice sheet's growth compares with the ice it's losing from melting, or from pieces that break off and float away as icebergs. The amount of snow being added each year is just as important as the amount of ice that's lost in determining whether a glacier, ice cap, or ice sheet causes sea levels to rise.

Ice Chips

Scientists estimate snow falling on all the world's glaciers, ice caps, and ice sheets would decrease global sea levels by a quarter of an inch (6.5 millimeters) a year if the oceans were not gaining or losing water in any other way. Currently, sea level is rising by about 0.07 inches (1.8 millimeters) a year. Only a small imbalance in the amount of snow falling, or ice melting would greatly decrease or accelerate sea level rise.

In other words, an important aspect of understanding glaciers, ice caps, and ice sheets is figuring out whether the amount of snow coming into the system matches the amount of ice leaving. The earth goes into an ice age when the amount of snow piling up and not melting exceeds the amount melting or falling into the sea. Earth's ice has been decreasing—but not always at a steady rate—since the peak of the last ice age around 20,000 years ago. One question scientists are trying to answer today is whether the balance is shifting toward a faster decrease in ice and a consequent faster rise in sea level. If this is happening, we need to ask whether human actions, such as adding heat-retaining gasses to the air, are responsible, or are the changes all natural?

Big Icebergs, Ice Shelf Collapse Raise Questions

Antarctica's ice began making worldwide news in March 2000 when an iceberg about as wide as Long Island, but longer, broke away from the Ross Ice Shelf to be given the name "B-15" by the U.S. National Ice Center. A few days later, B-15 broke into two parts, thus becoming B-15A and B-15B. The next month, three more icebergs, which were from 22 to 63 miles (35 to 101 kilometers) long, broke off. Some of the news stories said that global warming was at work, and implied Antarctica was melting.

Explorations

The National Ice Center in Suitland, Maryland, which is operated by the U.S. Navy, the U.S. Coast Guard, and the National Oceanic and Atmospheric Administration, tracks and reports on sea ice and icebergs in the Arctic, the Antarctic, the Great Lakes, and Chesapeake Bay. The center tracks sea ice and icebergs with satellite images. Antarctic icebergs more than 11.5 miles (18.5 kilometers) long are named using a letter based on the quadrant they are first spotted in, and a number representing the number of icebergs named in that quadrant since the 1976, when the center began tracking bergs.

Then between January 31 and March 7, 2002, the 1,255 square mile (3,250 square kilometers) Larsen B Ice Shelf on the Antarctic Peninsula disintegrated, leaving thousands of icebergs floating in an area the size of Rhode Island that had been solid ice. "We were astonished when we saw whales, penguins and seals in the place where for thousands of years before there was 250 meters (800 feet) of ice," said Pedro Skvarca of the Argentine Antarctic Institute.

Big Icebergs Part of a Natural Cycle

Scientists who study Antarctica's ice generally agree that the huge icebergs that break off from the large ice shelves are part of the natural cycle Antarctica has seen since its ice sheets formed. In fact, the iceberg calving that started in 2000 brought the Ross Ice Shelf to about the size it was in 1911, when Robert Falcon Scott's British expedition first mapped it.

One reason scientists are sure warming isn't causing huge icebergs like B-15 is that even with the increase in global average temperatures, the Ross Ice Shelf, much less the rest of the main part of the continent, has not come close to warming above freezing. In fact, the few long-term records available show that main part of Antarctica might even have cooled a little since early in the twentieth century.

Warming Did Help Break Up the Larsen Shelf

Although there's no reason to think that warming has anything to do with creating Antarctica's big icebergs, scientists say warming was responsible for the collapse of the Larsen B Ice Shelf, but wonder whether it's connected to earth's general warming.

Unlike the rest of the continent, the Antarctic Peninsula has warmed by about 4°F (2.2°C) since the 1950s. And the December 2001 through February 2002 Antarctic summer seems to have been the warmest at the northern end of the Peninsula and nearby islands since records began early in the nineteenth century.

Scientists Have Some Ideas About the Collapse

Even though the Peninsula is warming, computer models of climate do not create the warming seen in the Peninsula, John King of the British Antarctic Survey told a scientific conference at Hamilton College, New York, in 2002. John Domack of Hamilton said the Southern Ocean has been warming. The Peninsula juts out into the ocean current that circles Antarctica. Maybe, Domack said, the Peninsula is warming quickly because "it is the finger that is stuck into the warming kettle."

Ice Chips

Does what happened to Larsen B show what's in store for the rest of Antarctica? "It would take a warming trend as extreme as the one in the Antarctic Peninsula for at least 50 years to bring the Ross Ice Shelf to the threshold of breaking up," says Ted Scambos of the National Snow and Ice Data Center in Boulder, Colorado. "We see no evidence of this."

Scientists Look at the West Antarctic Ice Sheet

Charles R. Bentley, now retired from the University of Wisconsin, and Robert A. Bindschadler, a senior fellow at NASA's Goddard Space Flight Center, wrote in the December 2002 issue of *Scientific American*, that theories about a possible collapse of the West Antarctic Ice Sheet "oversimplified the ice sheet's own dynamics, which so far have exerted enough control … to avoid, or at least forestall, a swift demise." They note that the Ice Sheet managed to hang around at the end of the last ice age when ice sheets covering large parts of Europe and North America were quickly—in geologic terms—disappearing.

Bindschadler and Bentley concluded "cautiously" that the ice sheet will continue shrinking slowly, adding another 3 feet (1 meter) to the global sea level every 500 years. But "before anyone breathes a sigh of relief, we must remember that this remarkable ice sheet has been surprising researchers for more than 30 years—and could have more shocks in store."

Explorations

A geological study of mountain peaks once buried by the West Antarctic Ice Sheet shows "there was a gradual and continuous melting," says John O. Stone, in the January 2, 2003, issue of *Science*. The ice has retreated by about two inches a year in a steady pattern that shows no sign of slowing. The study by Stone and his colleagues added to the evidence that the earth's warming is not speeding up the melting of the West Antarctic Ice Sheet, at least not yet.

Researchers Wonder About Greenland's Ice

Recent measurements show that Greenland's ice is melting quicker around the edges than it had been, but is probably growing in the center. This fits what you would expect with a warmer climate. While the warmth helps melt ice quicker, the temperature has to climb above freezing, which rarely happens across the 9,000-foot (2,740-meter) elevation in the middle of the ice sheet. Warming brings more snow because warmer air carries more of the water vapor needed to make snow crystals.

Jay Zwally of NASA's Goddard Space Flight Center and his colleagues reported in 2002 that they had found that the flow near the edge of the Greenland Ice Sheet speeds up in summer. In the places they measured, the flow speeded up from 12.3 inches (31.3 centimeters) per day in the winter to a peak of 15.7 inches (40 centimeters) a day in the summer. This happens because water from ponds formed by melted ice on top of the ice flows through holes in the ice to lubricate the bedrock at the bottom. "This process was known for decades to enhance the flow of small mountain glaciers, but was not known to occur in the large ice sheets," Zwally said.

Ice Chips

One reason to worry about Greenland melting before Antarctica in a warming world: It's done it before. Greenland ice cores show that during Earth's last warm period—110,000 to 130,000 years ago—most of Greenland's ice melted. Scientists estimated that much of the global sea level rise during that warm period came from melted Greenland ice.

As with Antarctica, scientists have a lot of questions about what's happening to Greenland's ice. While a general warming caused by humans is certainly one of the major suspects behind what seems to be a loss of ice, other things could be at work. These include the following:

- A speeding up of Atlantic Ocean currents that brings warm water north to around Greenland.

- A shift in air pressure and wind patterns connected with the North Atlantic Oscillation, which changes over decades (which we looked at in Chapter 25).
- The continued warming of the earth in the 10,000 years since the last ice age.

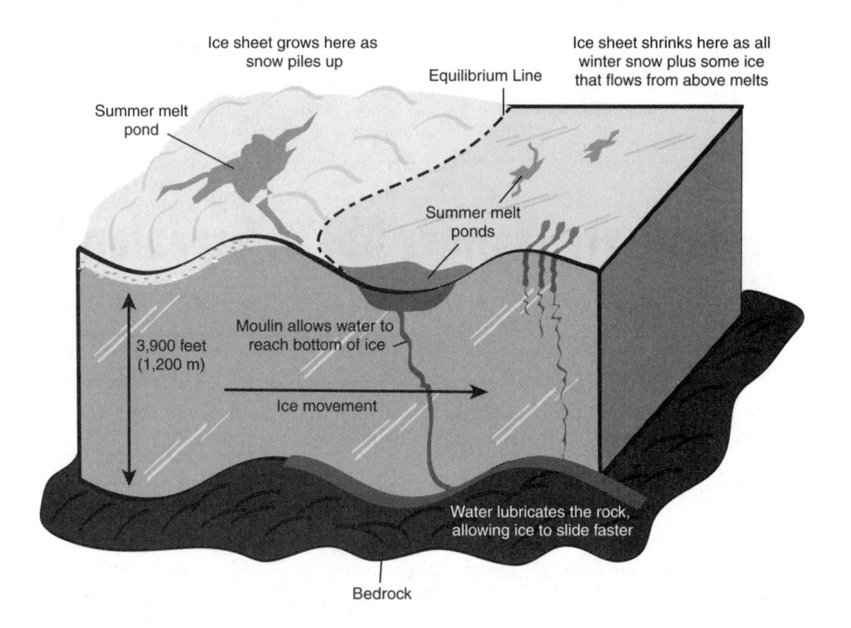

A look inside the edge of Greenland's ice sheet.

(Information from NASA)

Sea Ice Also Seems to Be Changing

Unlike ice that's sitting on land like Antarctica's two ice sheets and the Greenland Ice Sheet, sea levels don't rise when sea ice melts. This doesn't mean that we shouldn't be concerned about a possible loss of Southern Ocean or Arctic Ocean sea ice.

First, sea ice is an important part of both oceans' environment. A decrease would surely harm some of the creatures that live in the ocean, but could give others an advantage. Sea ice is a big player in the global climate system, although the details of exactly how it works aren't fully understood. But if the Arctic or Southern oceans ended up with less sea ice in the summer, the oceans would absorb solar energy that ice now reflects away. In the winter, the oceans would add more heat to the air than they do now. The Polar Regions wouldn't do as good a job as Earth's air conditioners.

Sea Ice Does a Good Job of Hiding Its Secrets

Determining what is happening with sea ice is much more difficult than measuring ice sheets, which is difficult enough. But at least the ice sheets stay still while you measure them. Huge amounts of sea ice begin forming each fall and continue growing until the end of the winter, when it all begins melting.

We can compare how much of the Southern Ocean or Arctic Ocean is covered by ice near the end of winter when it normally covers the largest area, and in late summer when the oceans have the least sea ice. But the amount of ice and the parts of the oceans it covers changes from year to year. The lack of good, long-term records makes sea ice studies difficult. If you want to see if air temperatures are changing, you can look up weather records for large parts of the world going back more than a century.

No real records of sea ice exist for the years before satellites began capturing ice images. Researchers can get some ideas about which parts of an ocean sea ice covered or didn't cover at particular times by looking into historical accounts. But such records only hint at the extent of past sea ice.

Sea Ice Increasing in Parts of Southern Ocean

In 2002, Claire Parkinson of NASA's Goddard Space Flight Center published her analysis of the length of the Southern Ocean sea ice season, using satellite records for 1979 through 1999. Her conclusion: In some areas the number of days with sea ice have increased, while in other areas, the number of days with sea ice have decreased. The areas with a longer sea ice season cover about twice as much of the ocean as the areas with a shorter season. "You can see … that what is happening in the Antarctic is not what would be expected from a straightforward global warming scenario, but a much more complicated set of events," she said.

The patterns are complicated, but in general the largest decreases are near the northern end of the Antarctic Peninsula, which has seen Antarctica's only temperature increase since the 1950s. The Ross Sea area has seen the largest increases.

Arctic Ocean Sea Ice Is Decreasing

Unlike in the Southern Ocean, researchers generally agree that the total amount of sea ice in the Arctic Ocean is decreasing, but they disagree about what's going on.

Researchers from the National Snow and Ice Data Center in Boulder, Colorado, reported in December 2001 that the Arctic Ocean's sea ice cover in September 2001 was the least ever seen. That month, satellites showed that about 2 million square miles (5.2 million square kilometers) of the Arctic Ocean was covered by ice, compared with the average of about 2.4 million square miles (6.2 million square kilometers), since satellite monitoring began in the 1970s.

Scientists are also interested in whether the Arctic Ocean's ice is growing thinner. During the Cold War, U.S. submarines roaming under the Arctic Ocean's sea ice regularly measured its thickness—handy information to have if they had ever needed to launch missiles at the Soviet Union. With the end of the Cold War, this data became available. When compared with new measurements, it showed that the average thickness of Arctic ice in the areas measured decreased by more than 40 percent from the late 1950s to the mid-1990s.

Most researchers who study the Arctic Ocean suspect that human-caused global warming is at least part of the cause. But they also think the Arctic Oscillation, or maybe other natural swings in atmosphere-ocean patterns are also partly responsible. Some think that some of what looks like a loss of ice thickness could be a matter of winds connected with the Arctic Oscillation blowing the ice into different parts of the ocean.

Researchers Hope for Help Finding Answers

In January 2001, the Intergovernmental Panel on Climate Change (IPPC) said its best estimate was that, during the twenty-first century, Antarctica would end up locking up more water than it releases through melting of its ice. While warmer air is expected to bring more snow to the continent, temperatures there are expected to stay below freezing. Antarctica could end up stashing some of the water melting in Greenland adds to the oceans.

Those who study the Antarctic and Greenland ice sheets have high hopes that NASA's new ICESat satellite, which is designed to produce precise measurements of ice sheet elevations, will begin giving them the continuous data from both Greenland and Antarctica they need to track the ice.

Figuring out what the atmosphere, oceans, and Antarctic ice have done in the past and might do in the future involves pulling together clues from a host of sources ranging from fossils buried in ocean-bottom mud to the outputs of more and more powerful computer models. It's a complicated detective story, Eugene Domack of Hamilton College says. "No one page is going to give you the whole plot; you have to read the book."

The Least You Need to Know

- If all the ice covering Greenland and Antarctica melted, the world's oceans would rise more than 250 feet (76 meters).
- Warming caused the 2002 collapse of an Antarctic Peninsula ice shelf, but not big icebergs elsewhere.

- Scientists today don't think the West Antarctic Ice Sheet is in as much danger of collapsing as they once did.
- Melting of the Greenland Ice Sheet seems to be speeding up.
- Some parts of the ocean around Antarctica seem to be gaining sea ice, while others are losing it.
- Researchers hope a new satellite will help them better measure what the ice is doing.

Chapter 28

Global Warming and the Polar Regions

In This Chapter

- What we mean by global warming
- The Northwest Passage could open by 2050
- Decreasing Arctic ice will change the ocean's life
- Changes in ocean currents could complicate climate
- Changes in Antarctica won't be as obvious as in the Arctic

In this chapter, we will begin with a look at what scientists mean when they talk of "global warming" and move on to how warming and the climate changes it brings are affecting the Arctic and Antarctic. We will focus more on the Arctic than the Antarctic because all indications are that Antarctica is less likely to see major changes in the coming 50 years or so than the Arctic. Still, the Antarctic is being affected by climate change and is an important part of the global climate system.

Global Warming and Climate Change

Since the 1980s, we've heard a lot about "global warming," and this term has become a synonym for *climate change*, which is unfortunate because climate change involves a lot more than the earth growing warmer. The important point is that the factors, whether natural or caused by humans, that are warming the earth on the average, are forcing the climate to react in many ways. The earth and its climate are so complex that no matter how much it changes, cold or mild winters as well as hot or comfortable summers, will continue. The changes will surely include more rain in some places, droughts in others, some currently barren areas could become farmland, and some farms could become wastelands. We could even see parts of the globe turning colder.

Polar Talk

Climate change is a change in the climate, especially of the normal temperature or precipitation patterns, of a place or region that would have important environmental, economic, or social effects. Climate change can be caused by human actions, natural causes, or a combination of the two.

Global Change Making Itself Felt in Polar Regions

As we saw in Chapter 27, we are seeing changes in polar ice, and at least some of this change is surely part of the more general, worldwide climate change. In Antarctica, even large events such as the collapse of the Larsen B ice shelf affect few people. This is not true in the Arctic, where most residents seem convinced they are seeing climate change. Some of the changes could be the weather's normal ups and downs, some could be the results of normal decade-to-decade weather swings, and the gases that humans are adding to the atmosphere could be causing some. Climate scientists face the challenge of sorting all this out.

Ice Chips

From 1971 to 2000, the annual average temperature in Barrow, Alaska, increased by 4.16°F (2.31°C) to 10.43°F (–11.98°C), according to the Alaska Climate Research Center at the University of Alaska, Fairbanks. Spring saw the biggest seasonal increase, 6.97°F (3.86°C).

Many of those who have lived in the Arctic all their lives also talk about the climate changing. For example, Inupiat Eskimo elders from Barrow, Alaska, say the ice is melting earlier in the spring and sea ice conditions are harder to predict than they used to be. Traditional cellars dug into the permafrost are beginning to thaw, allowing caribou and whale meat to spoil.

The changes that people are noticing around the Arctic Ocean go beyond warming, and illustrate why "climate change" is a better term than "global warming." For instance, the decrease in Arctic Ocean sea ice combined with changing wind patterns, which are apparently part of the Arctic Oscillation that we looked at in Chapter 25, are causing increased erosion from storms on Alaska's Arctic Coast. In April 2002, U.S. Republican Senator Ted Stevens of Alaska said: "Pack ice, which insulates our coastal villages from winter storms, has shrunk. Increased storm activity has caused significant beach erosion that may displace entire communities along the coastline of Alaska."

Caleb Pungowiyi, a former president of the Inuit Circumpolar Conference, said one of his aunts, who was born in 1912, once told him, "The world is faster now." What she meant, Pungowiyi said, was "back in the old days they could predict the weather by observing the stars, the sky, and other events." They can't do this any longer. "The weather patterns are changing so quickly she could think the earth is moving faster now."

The Arctic Research Consortium of the United States took her statement, *The Earth Is Faster Now,* as the title of a book about the growing effort by scientists to use indigenous observations of the Arctic environment. While to most people elsewhere, climate change might be a topic worth a few minutes of idle talk, for those who live around the Arctic Ocean, it's a common topic of conversation because people believe they are seeing climate change going on around them nearly every day.

Explorations

Inuit traditions include great respect for elders. Climate change, however, is eroding this respect, elders told researchers at workshops on global warming held across Arctic Canada in late 2002 and early 2003. This is happening because change is making the elders' traditional knowledge unreliable, Canada's *Nunatsiaq News* reported. Chris Furgal of Laval University, who headed the project, said elders see part of their roles in the community as no longer important. As one elder said: "Before we knew by looking at the sky whether there would be storms or if it would be calm. Nowadays just when you think you know how the weather will be, it can change in an instant. It's this inconsistency that is most noticeable."

The Northwest Passage Could Open by Mid-Century

Since ships first sailed both the Northeast and Northwest Passages in the late nineteenth and early twentieth centuries (see Chapter 11), few ships have sailed either passage, and neither has become the dreamed of commercial route. Now, however, the Arctic Ocean's decreasing ice is prompting a new, serious look at the possibility that by the middle of the twenty-first century, ships will be able to sail across the Arctic Ocean in the summer.

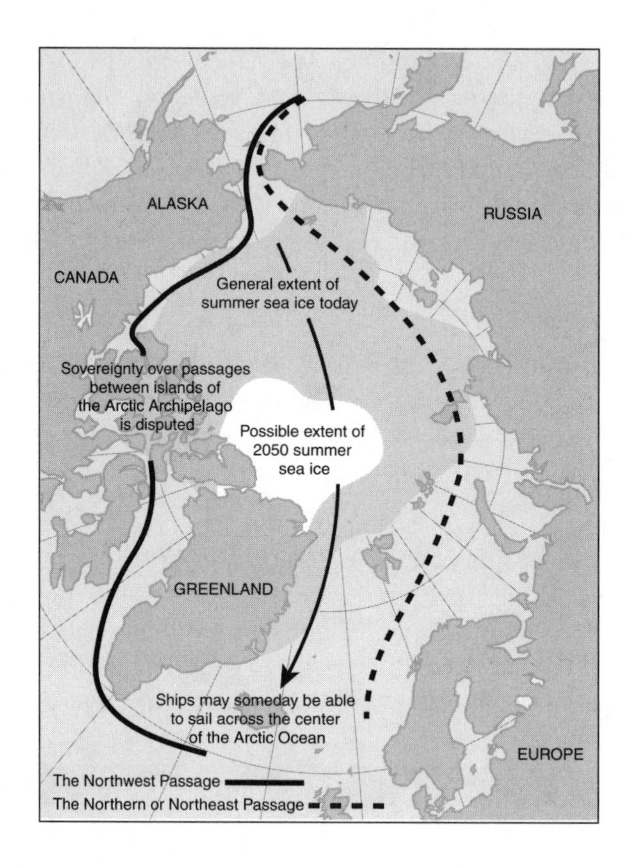

Possible Arctic Ocean ship routes.

More to It Than a Shorter Sea Route

Today ships sailing between Europe and Asia via the Panama Canal travel approximately 14,500 miles (23,300 kilometers). An open Northwest Passage would cut the distance to around 9,000 miles (14,500 kilometers). Even if the passage were open for a few months in the summer, however, icebergs would continue to be a danger and ice-strengthened ships would almost surely be needed.

Of course, the Northwest Passage is open in a sense today, but not really for commercial shipping. You can book a cruise through the Northwest Passage, or—for that matter—to the North Pole. Sending more ships, through the Passage, especially oil tankers, would greatly increase the danger of oil spills, which would be even harder to clean up and cause more environmental damage than in other oceans.

Ships Would Navigate a Legal Tangle

Canada claims sovereignty over the waterways between the passages of the islands of the Arctic Archipelago that Northwest Passage ships use. Most sea powers, however, including the United States, disagree, arguing that the Arctic Archipelago passages are like the Straits of Malacca between Malaysia and the Indonesian island of Sumatra, that connect the Indian and Pacific Oceans, and which are considered international waters.

So far, with a couple exceptions, these disagreements have been theoretical. In 1969, the United States made its point by sending an experimental, ice-strengthened oil tanker, the *Manhattan*, through the Northwest Passage without asking Canadian permission to sail through the Archipelago.

Ice-Free Ocean Would Change Arctic Life

While computer models of the climate see a continuing decrease in Arctic summer sea ice, most of them, but not all, see ice returning to the Arctic Ocean in the winter.

Even a reduction in summer sea ice would affect the life of the Arctic Ocean because the ice is one key part of the ocean's food web. At the other end of the Arctic food web, both Inuit hunters and polar bears depend on sea ice to stalk seals and walruses, which, in turn, require ice to give birth.

Ice Chips

Without sea ice, polar bears can be stranded on land with little to eat. In late July 2002, winds pushed sea ice up to the Barrow, Alaska, beach, stranding some polar bears until ice returned in the fall. The two men at the NOAA observatory near Barrow posted on their website a photo of three polar bears and noted the bears were waiting for the "two-legged seals" (the scientists) "to come out of their wooden den."

An almost complete disappearance of summer sea ice in the Arctic Ocean wouldn't end life there, but it would certainly change it. In fact, some kinds of life are likely to do well in the new conditions. One projection, for instance, is that some kinds of fish, such as Arctic cod, would become more abundant and that with less ice to contend with, commercial fishing in parts of the Arctic Ocean could become a major industry.

One of the goals of the Western Arctic Shelf-Basin Interactions Project, which we looked at in Chapter 21, is to collect the data needed to make better projections of how changes in Arctic Ocean sea ice will affect all of the ocean's living things, including the people who have depended on the ocean for food for centuries.

Some Fear Warming Could Slow Ocean Currents

One of the possibilities of a warming climate is that added rain falling on the northern Atlantic Ocean, fresh water from melting ice, or changes in the circulation of the Arctic Ocean could slow down the global oceanic conveyor belt that moves warm and cold water around the globe, both on top of and deep under the oceans.

The possibility of a slowdown in this thermohaline circulation means that global warming world could actually cool Europe. The Gulf Stream and North Atlantic Drift would bring less warm water into the North Atlantic. Europe is much warmer today than similar North American latitudes because winds from the west carry northern Atlantic warmth over the British Isles and the continent.

If changes in the North Atlantic managed to shut down the thermohaline circulation, the potential for cooling could extend beyond Europe, but no one really knows. Ice cores show that as the last ice age was ending, the climate saw abrupt swings between warm and cold periods over only a few years. Changes in the thermohaline circulation are among the leading suspects as the cause of these changes. Many climate scientists think abrupt climate changes could be a bigger challenge to human societies than a steadily warming climate.

Ice Chips

Some scientists argue that Earth's wild climate changes more than 10,000 years ago helped humans evolve into creatures with huge brains (compared with other animals) that worked together to hunt large animals. When the climate calmed down around when the ice age ended, humans were ready to develop agriculture and civilizations, taking advantage of the more benign climate we've enjoyed since then.

Most theories about a possible slowing or even stopping of the thermohaline circulation are based on the idea that the polar regions, and the oceans directly affected by the polar regions, are the main drivers of the circulation. But as with many other aspects of the earth's climate, much remains to be learned about the thermohaline circulation and it's possible that events in other parts of the world, such as hurricanes, could play a role.

In 2001, the Intergovernmental Panel on Climate Change said most climate models show the thermohaline circulation weakening, but Europe should continue to warm up. "However," the report says, "it is too early to say with confidence whether an irreversible collapse in the thermohaline circulation is likely or not and at what threshold it might occur and what the climate implications could be." In other words: Stay tuned.

The Antarctic Is Less Likely to See Big Changes

Some day Antarctica's large ice shelves could go the way of the Larsen B ice shelf, but most who study Antarctica's ice, along with the Intergovernmental Panel on Climate Change (IPCC), don't see this happening in the twenty-first century. In its 2001 report, the IPCC notes that the West Antarctic Ice Sheet has been the biggest concern because its bottom is below sea level. Yet the IPCC says current computer models project that the West Antarctic Ice Sheet will not add more than .117 inch (3 millimeters) a year to global sea levels over the next thousand years.

Essentially, Antarctica is a huge chuck of ice that will need a long time to warm to its melting point even if the rest of the world warms fairly quickly.

In Chapter 25, we looked at the links between the Antarctic and global climate, and saw how the continent and the Southern Ocean and its sea ice can make themselves felt far away. Researchers have learned that the Ross Sea area of Antarctica has a disproportionate influence on weather away from the continent.

This is one of the reasons the National Science Foundation is supporting the Ross Island Meteorological Experiment (RIME), which aims to learn more about the weather in this part of Antarctica and how it influences global weather and climate. Scientists involved expect that, by the time they spend the 2005–2006 and 2007–2008 Antarctic summers intensively measuring weather in the Ross Island area, and another couple years analyzing the results, they'll have a better understanding of how Antarctica and the rest of the world interact.

Sea Level Change Is the Big Question

We've seen how the polar regions are the earth's air conditioners that interact with the rest of the earth in many ways. As humans add carbon dioxide and other greenhouse gases to the atmosphere, global climate is being affected, and some of the biggest changes are likely in the polar regions, especially around the Arctic. The big question, however, is how much the level of the oceans around the world is going to rise.

Large amounts of polar ice are not going to melt in the lifetime of anyone living now. But the 2001 IPCC report does project that sea level will increase during the twenty-first century, with the smallest likely rise being 3.48 inches (.29 meters) and the largest being 2 feet, 11 inches (.88 meters). Around half of any increase would be from the "thermal expansion" of the water—water like other materials expands when it warms. Most of the rest would come from glaciers and small ice caps and other sources.

> **Explorations**
>
> The plot of the 1995 movie *Waterworld* was based on a complete melting of polar ice drowning most land. Melting all of Antarctica's ice would take average global temperatures climbing about 36°F (20°C) higher than today's temperatures, the IPCC says in its 2001 report. This is more than three times the greatest warming seen as possible in the twenty-first century, and is "a situation that has not occurred for at least 15 million years and which is far more than predicted by any scenario of climate change currently under consideration." If all of Earth's ice melted, sea level would rise by about 262 feet (80 meters), the U.S. Geological Survey estimates. Even if you doubled this to account for the expansion of heated seawater, huge areas of the earth would stay dry.

Greenland could lock up more water than it releases while Antarctica is almost sure to lock up more water than it releases. While the amount of sea-level rise doesn't sound like much, it could wipe out some low-lying areas, especially islands. And even slightly higher seas mean that water pushed ashore by hurricanes and other storms will go farther inland, doing more damage.

Polar Regions Will Remain Places for Awe, Adventure

While we don't have to worry about a "waterworld," we can be sure that over the coming years we will be seeing changes in the polar regions, with the Arctic seeing the biggest changes. Those most likely to be directly affected are the indigenous people around the Arctic whose traditions and subsistence are bound up with hunting sea mammals. These people can only hope that the loss of Arctic Ocean ice will slow down—that it's more a result of climate patterns such as the Arctic Oscillation, which will swing back to a mode that brings more ice. Many who are knowledgeable about the ways of climate fear, however, that climate change, maybe acting in part through the Arctic Oscillation, will continue warming the Arctic.

Another fear that's very much alive among those who live around the Arctic Ocean is that the combination of higher seas and bigger storm waves in an ocean with less ice will destroy their coastal villages. For instance, people in Barrow, Alaska, and other communities expect the ocean to take parts or all of their towns.

Even if this is happening, we will still be able to travel to the Arctic and Antarctic on cruises, for adventure trips such as skiing the last degree of latitude to the North Pole, or to sign up for one of the many jobs that can take you to the ends of the earth. None of us, however, will be able to really live the adventures of the past, whether its Amundsen and his men being the first to reach the South Pole, or Sir John Franklin and his men dying in the Arctic. The real polar adventures of today and tomorrow are the intellectual adventures of unraveling the secrets of how the Arctic, the Antarctic, their ice, water, land, and living things are all a part of our larger world. Maybe some of you will join the men and women who are setting off on scientific adventures in the polar regions. Whether you do this or not, we hope this book will help you become a connoisseur of past, present, and future polar adventures.

The Least You Need to Know

- "Global warming" really refers to many climate changes.
- The polar regions are feeling climate change more than other parts of the world.
- People who live around the Arctic are seeing signs of climate change.
- Melting Antarctic ice isn't expected to add much to sea levels during the twenty-first century.
- What may seem like small rises in sea level could have disastrous effects, including around the Arctic.

Appendix A

Glossary

ablation Removal of material from a glacier or ice sheet by melting, evaporation, or calving.

active layer The soil above permafrost that melts each year in the summer. Plant roots and burrowing insects are found here.

airship A lighter-than-air aircraft with an engine that moves it through the air, and which can be steered. The main body of a typical airship is a huge, cigar-shape balloon filled with a lighter-than-air gas, usually helium today.

alloy A homogeneous mixture of two or more metals, the atoms of one replacing or occupying positions between the atoms of the other. Brass is an alloy of copper and zinc.

altitude sickness A variety of symptoms resulting from body not obtaining the oxygen it needs in lower-pressure air at high altitudes. Effects vary widely and the condition can be deadly.

anhydrobiosis A state of apparent suspended animation that some invertebrates enter to survive dryness or other extreme stresses. Animals in this state appear to be dead for days, weeks, or even years until moisture returns, when they come back to life.

Antarctic The region of the earth between the Antarctic Circle and the South Pole.

Antarctic, Arctic Oscillations Patterns of air pressures and winds centered over the Antarctic or Arctic with two phases. The positive phase is characterized by high air pressure and strong winds ringing the polar region. The negative phase has lower air pressure and weaker winds around the polar region.

Antarctic Bottom Water Cold, very salty water that forms around Antarctica and flows along the bottoms of the world's oceans.

Antarctic Circle Latitude 66 degrees, 30 minutes south.

Antarctic Circumpolar Current The world's largest ocean current, which moves all around Antarctica. It connects the waters of the Atlantic, Indian, and Pacific oceans.

Antarctic Convergence The zone between about 50 and 60 degrees south where cold water from the Antarctic region meets and sinks beneath warm water from the middle latitudes. Sometimes called the "Antarctic Polar Front."

Antarctic Divergence The oceanic zone just north of the Antarctic continent where west-flowing and east-flowing currents diverge, allowing salty, nutrient-rich water to come to the surface.

Antarctic Polar Front *See* Antarctic Convergence.

Antarctica The continent that is almost entirely south of the Antarctic Circle.

Arctic The region of the earth between the Arctic Circle and the North Pole.

Arctic Circle Latitude 66 degrees, 30 minutes north.

astrobiology The multidisciplinary study of the possible origin, evolution, distribution, and future of life, and the likely limits on life elsewhere in the universe, using methods and insights of both the biological and physical sciences. Includes the study of life in extreme environments on Earth.

astronomical twilight The time before the sun rises or after it sets when the sun's center is 12 to 18 degrees below the horizon. Some sky is illuminated by the sun in this position.

atom The smallest unit of an element, having all the characteristics of that element and consisting of a dense, central, positively charged nucleus surrounded by electrons.

aurora australis The luminous phenomenon in the Southern Hemisphere created by energetic particles from the sun hitting atoms of atmospheric gasses.

aurora borealis The luminous phenomenon in the Northern Hemisphere created by energetic particles from the sun hitting atoms of atmospheric gasses.

auroral oval An oval band around each geomagnetic pole where organized auroras are located.

biomass The total mass of a particular organism or group of organisms living in a location or the entire world.

biosphere The part of the earth, including the oceans and the atmosphere, where life is found. Used by climate scientists to refer to living things as they are affected by the atmosphere and in turn affect the atmosphere.

blizzard In the United States, a severe winter storm with low temperatures, winds of 35 mph or faster, blowing snow that reduces visibility to 0.25 mile (400 meters) or less, lasting at least 3 hours.

boomeranged To be on a flight that lands at the airport it took off from. In Antarctica, this usually occurs because of an unforecast change to dangerous weather before the airplane reaches the "point of safe return," when it has enough fuel to return.

breathing-hole hunting A seal makes many holes at which to breathe. A harpoon hunter waits at one and strikes if the seal surfaces there.

calving The formation of an iceberg as ice breaks off a glacier or ice shelf.

carnivore A flesh-eating animal.

CFC Abbreviation for chlorofluorocarbons, synthetic chemical compounds of chlorine, fluorine, and carbon. Used as refrigerants and solvents, among other uses. They are nonpoisonous and nonflammable, but in the upper atmosphere ultraviolet energy breaks them down and their chlorine atoms destroy ozone.

Circumpolar Deep Water North Atlantic Deep Water that has been modified by interactions with other water masses on the way to Antarctica, where some of it upwells.

civil twilight The period before the sun rises or after it sets when the sun's center is less than 6 degrees below the horizon. It's usually light enough to engage in outdoor activities without artificial light.

climate Informally, the average weather for a particular place or region. Meteorologists usually use statistical information averaged over 30 years to give "normal" temperatures, precipitation, and other factors plus information on weather extremes to define a place's climate.

climate change A change in the climate, especially of the normal temperature or precipitation patterns, of a place or region that would have important environmental, economic, or social effects. Could be caused by human actions, natural causes, or a combination of the two.

compound bow A bow that uses a system of cables and pulleys to make the bow easier to draw.

copepods Microscopic to nearly microscopic crustaceans that are a common kind of zooplankton.

cosmic rays High-energy particles that move through space at close to the speed of light and bombard Earth from all directions.

crampon Spiked plates that fit onto shoes to prevent slipping on ice.

cyanobacteria Bacteria that live in water and which have the ability to create plant matter with photosynthesis. They are single cells, but often grow in colonies that can be seen with the unaided eye. They first appeared on Earth more than 3.5 billion years ago, making them the oldest-known fossils, but remain an important part of life in water.

cryosphere The parts of the earth covered by ice for part or all of the year.

dark energy Unknown energy that might make up 65 percent of the universe. It shoves galaxies away from each other at an ever-increasing speed.

dark sector The area at the South Pole Station where many of the pole's astronomical instruments are located, half a mile from where the people live, to ensure no light (visible or infrared), contaminates observations.

DNA Abbreviation for deoxyribonucleic acid, which is a thin, chainlike molecule found in every living cell. It directs the formation, growth, and reproduction of cells and organisms. Short sections of DNA called genes determine heredity; that is, the passing on of characteristics, in living things.

electromagnet A magnet consisting of a coil of insulated wire wrapped around a soft iron core that is magnetized only when current flows through the wire.

electron A subatomic particle having a negative electric charge.

El Niño-Southern Oscillation (ENSO) Coupled ocean-atmospheric interactions in the tropical Pacific characterized by episodes of anomalously high sea surface temperatures in the tropical eastern Pacific; associated with large-scale swings in surface air pressure between the western and central tropical Pacific. The phase with warm water in the eastern Pacific is known as El Niño and the phase with cool water there is known as La Niña.

equinox The times of the year when the sun is directly over the equator, around March 20 and September 22.

Eskimo Word commonly used for native people in Arctic North America and Greenland. Today Canadian and Greenland natives consider it derogatory and use *Inuit* instead. Alaska's Arctic native people don't find the word offensive and use it more or less interchangeably with *Inupiat* to refer to themselves.

extratropical storm A large storm that forms over land or a cool ocean. The storm has fronts—boundaries between warm and cold air. Air temperature contrasts are the main energy source.

fast ice Sea ice that forms and remains attached to the shore, extending a few yards or a hundred miles from the coast. Fast ice is different from an ice shelf in that fast ice is sea ice while a shelf is made of ice that pushed over the ocean from land.

feedback In Earth sciences, a process where a change in a variable, through interactions in the system, either reinforces the original process (positive feedback) or suppresses the process (negative feedback).

firn Rounded, small grains of snow that are bonded together and which are at least a year old.

fossil A part of an organism (like a leaf) from a past geological age that's embedded and preserved as a rock in Earth's crust.

frostbite The freezing of the skin or tissues under the skin. If the tissue isn't warmed soon enough, it can die.

glacier A body of natural ice on land that flows or has moved in the past.

glacier ice (glacial ice) Well-bonded ice crystals compacted from snow.

Gondwanaland The proto continent of the Southern Hemisphere that broke up into India, Australia, Antarctica, Africa, and South America.

greenhouse effect Warming of the earth caused by greenhouse gases that absorb infrared energy earth radiates, which warms them. The warmed gases radiate energy in all directions, making the atmosphere warmer than it would otherwise be.

greenhouse gases Gases that absorb heat. Water vapor is the main greenhouse gas. Others are carbon dioxide, added to the air naturally, and by human activities such as burning fossil fuels.

halocline A layer of water in which the salinity increases rapidly with depth.

harpoon A spear like weapon with a barbed head used for hunting whales. The head is attached to a line the hunter holds and the head comes off the shaft upon impact with the animal.

haul out, hauled out In reference to a marine mammal such as a seal, to come out of the water, to be out of the water on ice or land.

heat budget An inventory of all the heat flowing into and out of a system and the heat stored in the system.

Herbie Name for a local windstorm on Ross Island, Antarctica. These storms can bring high winds, zero visibility, and dangerous wind chills to the McMurdo Station.

herbivore An animal that chiefly eats plants.

hoar frost A deposit of interlocking ice crystals formed by direct deposition of water vapor on objects, including snow or grass, freely exposed to the air.

Homo sapiens Modern species of human beings, the only surviving type. Arose in Africa about 150,000 years ago.

housemouse The person at a polar research camp who spends a day doing various domestic chores, such as cleaning the kitchen. This duty rotates among everyone in the camp.

hypothermia A decrease in the core body temperature that impairs normal muscular and cerebral functions. It occurs when the body loses heat faster than it is replaced. Symptoms begin when the core body temperature drops below 95°F (35°C).

ice age A period in the earth's history when ice sheets cover vast regions of land, typically lasting about 100,000 years.

icebreaker A ship designed to break through pack ice, with an extra-strong hull and more-powerful engines than ordinary ships of the same size. They break the ice by riding the front of the ship up onto the ice where the ship's weight breaks it.

ice cap A dome-shape glacier ice mass, spreading out in all directions, usually less than 19,300 square miles (50,000 square kilometers).

ice core A cylindrical piece of ice removed from an ice sheet, ice cap, or glacier by a drill designed to do this. The term can refer to all of the ice removed from a hole drilled to the bottom of an ice sheet, or to individual 3-foot (1-meter) pieces of this ice.

icefish Antarctic fish that are members of the family *Channichthyidae* with no hemoglobin, the oxygen-carrying pigment that makes red blood cells red.

ice floe Any contiguous piece of sea ice. Floes can be as small as 20 meters across and larger than 10 kilometers.

ice sheet A dome-shape mass of glacier ice that covers surrounding terrain and is greater than 19,300 square miles (50,000 square kilometers), such as the Greenland and Antarctic ice sheets.

ice-strengthened ship A ship with a strong hull designed to move through pack ice that isn't solid enough to require an icebreaker. The ship can withstand collisions with ice floes.

ice wedge A buildup of ice in frozen soil, that is wedge-shape in cross-section.

infrared (heat) astronomy The study of space radiation with wavelengths beyond the red end of the visible spectrum, from approximately 800 to 1 millimeter.

interglacial A period in Earth's history when ice sheets generally disappear, typically lasting 10,000 to 20,000 years. Earth has been in an interglacial for about 11,000 years.

Inuit A word that means "people" and refers to the present-day native peoples in Canada, Greenland, and sometimes Alaska.

inversion In meteorology, a departure from the usual decrease of temperature with altitude.

invertebrate An animal without a backbone or spine. Such animals have less well-developed brains than vertebrates.

isotopes Forms of atoms of an element that differ in atomic weight. All atoms consist of a nucleus made of protons, with a positive charge, and neutrons, with no charge, surrounded by a cloud of electrons, with negative charge. The number of protons and electrons are the same, but the number of neutrons can differ in atoms of an element. Oxygen, for example, has eight protons, but can have eight, nine, or ten neutrons, giving it three isotopes.

Jamesway A prefabricated, canvas building shaped like a half circle with a wooden floor and frame, which is easily erected and taken down. Many dating back to the time of Operation Deepfreeze I are still used at U.S. Antarctic bases and field camps.

JATO bottles Jet-Assisted Take Off rockets, to provide added thrust for airplanes taking off on short runways, at high elevations, or in snow. The name JATO goes back to World War II, but they are rockets, not jets. The 109th Airlift Wing, which uses them on its LC-130s calls them "ATO" for Assisted Take Off.

katabatic winds A wind created when cold, dense air begins flowing down hill.

krill The common name used for members of the crustacean order *Euphausiacea.* Five of the world's 80 plus species are found in the Southern Ocean.

land-based ice sheet The bottom of the ice sheet is generally above mean sea level.

latitude Imaginary lines circling the earth from east to west. Also called parallels because they are all parallel to each other. The equator is 0 degrees latitude, the North Pole 90 degrees north, and the South Pole 90 degrees south.

lead Any fracture or passageway through sea ice that is navigable by surface vessels.

lemming A mouse-size animal that looks like a miniature guinea pig, common in the Arctic.

Little Ice Age The general cooling of Europe and other parts of the Northern Hemisphere that began in the middle of the twelfth century and lasted until about 1850. The coldest period began around the middle of the fifteenth century.

longitude Imaginary lines circling the earth from the North Pole to the South Pole, at right angles to latitude lines. The 0 degrees longitude line runs through Greenwich, England, on the edge of London.

magnetic field A force field in the region around a magnet or an electric current, characterized by lines of magnetic force connecting the two poles of the magnet or surrounding the electric-current flow.

magnetic poles The points in the Arctic and Antarctic toward which the needles of magnetic compasses point. They are not at the geographic poles, which is why compass needles do not point due north and south from most places on Earth.

mantle The layer of Earth between the crust and the core.

magnetosphere A region surrounding Earth, extending from about 60 miles (100 kilometers) to several thousand miles above the surface, in which charged particles are trapped and their behavior is dominated by Earth's magnetic field.

meteorite A stony or metallic mass of matter that has fallen to the earth's surface from space.

microwave radiation A high-frequency electromagnetic wave, one millimeter to one meter in wavelength, between infrared (heat) and short-wave radio wavelengths.

midrats Military term for "midnight rations," for those who work at night. Often the rule is that night workers have priority over those who just want a midnight snack.

multi-year sea ice Ice that has survived at least two summers' melt. The ice is almost salt-free. Bare ice is usually blue.

marine-based ice sheet An ice sheet with the bottom generally below mean sea level.

marine mammal A mammal is an animal that gives birth to live young, nurses its young, and maintains a constant body temperature. A marine mammal depends on the sea for all or almost all its food.

Medieval Warm Period The time believed to have lasted from the eighth through the twelfth centuries when temperatures, especially in the Northern Hemisphere, were apparently a few degrees higher than during the preceding and following periods.

mesoscale In meteorology, phenomena that range in size from a few miles to around 60 miles (100 kilometers) across. These include thunderstorms, which do not occur in Antarctica, to local winds and small-scale storms, which are a big problem there.

metabolism The complex processes that enable an organism to maintain life.

middle latitudes The area of the earth between 23.5 degrees and 66.5 degrees north and 23.5 and 66.5 degrees south. The length of day changes noticeably during the year.

molt To shed feathers, hair, or skin periodically, often with the season, to allow new feathers, hair, or skin to grow.

musk ox A large, stocky ox that lives in the arctic.

Mustang Suit Insulated protective coveralls that keep someone who falls into very cold water alive long enough to be rescued.

nautical twilight The period before the sun rises or after it sets when the sun's center is 6 to 12 degrees below the horizon. At sea, the horizon is distinct.

neutrinos Elementary particles with zero charge and spin. They have very small mass and weak interactions with matter.

nilas A thin elastic crust of ice, easily bending on waves.

North Atlantic Deep Water A mass of dense, cold, salty water that forms on the surface of the northern Atlantic, primarily in the Norwegian Sea, and descends to flow south deep under the ocean.

Northeast Passage A route that ships could follow between the Atlantic and Pacific oceans by sailing along the ice-free Scandinavian Arctic Coast and Russia's Arctic Coast.

Northwest Passage A route that ships could follow between the Atlantic and Pacific oceans by sailing along the northern coasts of Canada and Alaska.

nunataks Tops of Antarctic mountains that stick up out of the Ice Sheet which covers most of the continent.

nutrients The many organic compounds from living things such as combination proteins and carbohydrates, and inorganic compounds, which are not from living things, such as nitrogen and phosphorous, which are necessary for plants and algae to grow.

pack ice Any area of sea ice, other than fast ice.

paleoclimatologist A scientist who studies climates of the past for which written weather records are not available, using data such as from ice cores.

pancake ice Mostly circular pieces of ice from 1 to 10 feet (30 centimeters to 3 meters) in diameter, and up to 4 inches (10 centimeters) thick, with raised rims caused by the pieces hitting one another.

Pangaea The super continent that broke up 200 million years ago to form the present landmasses.

permafrost Permanently frozen subsoil that maintains a temperature below 32°F (0°C) continuously for 2 years or longer.

pingo A hill created when water pools and freezes under the root mat. When the water freezes, it expands, pushing up the soil.

pinnipeds The suborder of marine mammals that includes seals, sea lions, and the walrus.

Pleistocene Epoch Geological period 1.6 million to 10,000 years ago. During that time, ice covered northern areas and then retreated.

polarization A state in which light-wave vibrations line up in the same direction.

polygon (on tundra) A pattern on the surface of frozen ground formed above ice wedges. They can have low or high centers.

polynya (pronounced *puh-LIN-yuh*) A large area of open water or reduced ice concentration in an ice pack.

ptarmigan A grouse that has feathered legs and feet for arctic life.

pterosaur An extinct flying reptile.

reverse osmosis A widely used method for desalting seawater. In normal osmosis, a less concentrated liquid, such as fresh water, will flow through a membrane into a more concentrated liquid, such as salt water. But applying enough pressure to the salt water will force it through a membrane, which filters out the salt.

rookery An area where large numbers of birds or mammals, such as penguins or seals, come together to breed or live.

Sami (Lapps) Nomadic herders who inhabit Lapland in the northern Scandinavian Peninsula, including parts of Norway, Sweden, Finland, and Russia.

sastrugi A long, wave-shape ridge of hard snow formed by the wind. They are common in the polar regions.

sea ice Any form of ice found at sea that has originated from the freezing of sea water.

sea ice ridge A line or wall of broken ice forced up by pressure.

sea ice keel The submerged, broken ice under a ridge, forced downward by pressure.

Seabees (CBs) Sailors who are skilled in construction trades and members of construction battalions. They first gained fame during World War II for building airfields and other facilities on captured Pacific islands while the bullets were still flying.

sedge Grasslike plants with solid stems, leaves in three vertical rows, and spikelets of tiny flowers.

shortwave radio Radios that transmit and receive in the 3 to 30 megahertz frequency range. Radio waves in this range are reflected by the layer of charged particles known as the ionosphere, more than 34 miles (55 kilometers) above the earth.

skiway A runway of groomed snow used by ski-equipped airplanes at bases and temporary research camps where planes will be coming and going.

Snowball Earth The theory that ice covered all, or almost all the earth, during four super ice ages, each lasting millions of years, from around 600 to maybe 800 million years ago. Geologists disagree about how extensive Earth's ice cover was during this time and whether thick ice ever covered even the tropical oceans.

solar flare A sudden eruption of hydrogen gas on the surface of the sun, usually associated with sunspots and accompanied by a burst of ultraviolet radiation that is often followed by a magnetic disturbance.

solstice The times at which the sun is above latitude 23.5 degrees north, around June 21, and latitude 23.5 degrees south, around December 21.

species Related organisms capable of interbreeding.

stratosphere The layer of the atmosphere that begins about 4 miles (6 kilometers) above the earth in the polar regions, and about 10 miles (16 kilometers) above the tropics. Air temperatures in the stratosphere do not decrease with altitude as in the lower atmosphere. The top of the stratosphere is about 31 miles (50 kilometers) up all over the earth.

sublimate, sublimation In meteorology, the change of state of water directly from ice to vapor without first melting into a liquid.

submillimeter astronomy A branch of astronomy that covers radiation wavelengths from approximately 0.3 to 1 millimeter. It is important in the study of giant molecular clouds, which are cool, dense regions of interstellar matter where atoms tend to combine into molecules.

sunspot Any of the relatively cool dark spots appearing periodically in groups on the surface of the sun that are associated with strong magnetic fields.

supernova An exploding star that can become billions of times as bright as the sun before gradually fading from view.

taiga (pronounced *ti-ga*) A Russian word for the thin, northern evergreen forest.

teleconnection A linkage between atmospheric or oceanic circulation changes occurring in widely separated regions of the world.

thermohaline circulation The vertical movement of ocean water driven by density differences caused by temperature and salinity variations. The term is derived from the Greek words for "heat" and "salt."

thermokarst lake A lake formed when water holds heat that thaws the permafrost below.

thermoregulation The ability of an animal to maintain body temperature within a narrow range despite changes in the environment. Animals that can do this are warm-blooded, animals that can't are cold-blooded.

traverse As used by polar and other scientists, refers to research using a party traveling across the land, or an ice sheet or ice cap. The party stops along the way to make scientific observations.

tree ring A sheath of cells forming a circle around a cross section of a tree limb or trunk. The cells are formed by each spring's growth spurt. Thus, they can be counted to determine the age of a tree.

tropics The region of the earth from latitude 23.5 degrees south (the Tropic of Capricorn) across the equator to 23.5 degrees north (the Tropic of Cancer) where the sun is directly overhead at least once a year.

tropical cyclone A large storm that forms over a warm ocean. The entire storm is warm with no fronts. Heat released as water vapor, condensing to form thunderstorms, is the main source of energy. Hurricanes and typhoons are tropical cyclones.

tundra A treeless area between the ocean or ice-covered areas and the tree line of polar regions, having permanently frozen subsoil and supporting low-growing vegetation such as lichens, mosses, and stunted shrubs. Alpine tundra is found high on mountains outside the polar regions.

Tunit Extinct Eskimos that occupied the Canadian and Greenland Arctic before the arrival of the Inuits. *Tunit* is the Inuit word for these folks.

umiak Large Eskimo boat made of skins stretched on a wooden or bone frame. Propelled by paddling.

ultraviolet light The invisible form of light with frequencies just higher than those of violet light. Ultraviolet energy causes sunburn and harms living matter in other ways.

upwell, upwelling The process by which deep, cold, nutrient-laden water comes to the ocean's surface, occurring in areas of divergence, such as around Antarctica or where winds blow water offshore, such as along the California Coast.

vascular plants Plants that have vessels to transport internal fluids. In addition to water-conducting tissues, most have true stems, roots, and leaves.

ventifacts Rocks in arid areas that have been eroded and polished by wind-blown sand and ice crystals.

vertebrate An animal with a segmented spinal column and a well-developed brain. Includes mammals, birds, reptiles, amphibians, and fish.

viscosity The property of a fluid that resists flowing. In most fluids, it increases as temperature falls.

volcanic bomb A semi-molten, streamlined fragment of compressed ash and pumice, which is ejected from a volcano.

Weasel A World War II tracked vehicle designed to haul soldiers and equipment over a variety of terrain, including snow. It was about 10 feet long and 5 feet wide, and weighed about 4,000 pounds (1,814 kilograms). Treads ran the vehicle's length on both sides.

whiteout A dangerous weather phenomenon that occurs when thick, low clouds reduce surface definition, and the horizon is obscured, making it difficult to know if you are on a flat or sloping surface, and to judge distances or sizes of objects.

Appendix B

The Earth's Poles

This list was compiled by R. K. Headland of the Scott Polar Research Institute at Cambridge University in England.

North Geographic Pole

- Location: Arctic Ocean at Latitude 90° 00'N, Longitude all.
- The Northern axis of Earth's rotation.
- First seen in 1926 from the airship *Norge*.

North Magnetic Pole

- Latitude 76° 06'N, Longitude 100° 00'W.
- The point that the north-seeking end of a compass needle points to.
- First reached by: Captain James Ross in 1831 when it was on the Boothia Peninsula in what is now Nunavut Territory, Canada.

North Geomagnetic Pole

- Latitude 78° 30'N, Longitude 69° 00'W.
- The Northern end of the Earth's geomagnetic field.
- First seen or reached by: unknown.

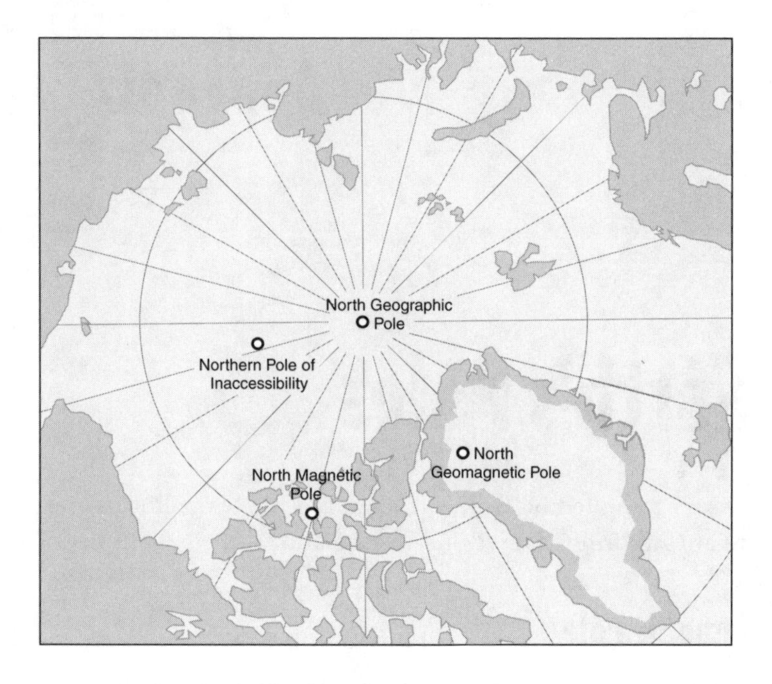

Northern Pole of Inaccessibility

- Latitude 84° 03'N, Longitude 174° 51'W.
- Location on the Arctic Ocean most distant from land.
- First reached by Sir Hubert Wilkins by aircraft in 1927.

South Geographic Pole

- Latitude 90° 00'S, Longitude all.
- The southern axis of Earth's rotation.
- First reached by Roald Amundsen's expedition in 1911.

South Magnetic Pole

- Latitude 65° 00'S, Longitude 139° 00'E in the Southern Ocean near Antarctica.
- The location the south-seeking end of a compass needle points to.
- First reached by a party led by Edgeworth David and including Alistair Mackey and Douglas Mawson in 1909 when the pole was inland in Antarctica.

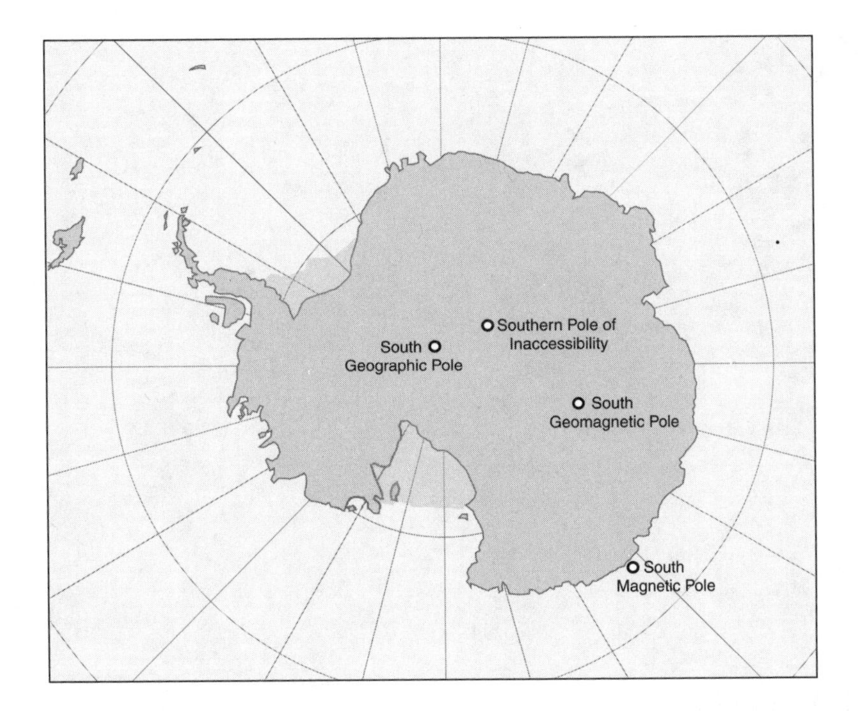

South Geomagnetic Pole

- Latitude 78° 30'S, Longitude 110° 00'E.
- The southern end of the Earth's geomagnetic field.
- First reached by men who in 1957 built the Soviet Union's nearby Vostok Station.

Southern Pole of Inaccessibility

- Latitude 85° 50'S, Longitude 65° 47'E.
- Location in Antarctica that is most distant from the Southern Ocean.
- First reached by the Soviet Expedition that established the Sovetskaya station in 1957.

Resources

The Arctic and Antarctic Today

Berkman, Paul Arthur. *Science Into Policy: Global Lessons from Antarctica*. San Diego: Academic Press, 2002.

Ehrlich, Gretel. *This Cold Heaven: Seven Seasons in Greenland*. New York: Pantheon Books, 2001.

Greenland Home Rule Government. *Greenland Atlas*. Nuuk, Greenland, 1993.

Hess, Bill. *Gift of the Whale: the Inupiat Bowhead Hunt, A Sacred Tradition*. Seattle: Sasquatch Books, 1999.

Krupnik, Igor, and Dyanna Jolly. *The Earth Is Faster Now*. Fairbanks, AL: Arctic Research Consortium of the United States, 2002.

Gorman, James. *Ocean Enough and Time: Discovering the Waters Around Antarctica*. New York: HarperCollins, 1995.

Hince, Bernadette. *The Antarctic Dictionary: A Complete Guide to Antarctic English*. Collingwood, Australia: CSIRO Publishing, 2000.

Lopez, Barry. *Arctic Dreams*. New York: Bantam, 1989.

Nielsen, Jerri, with Maryanne Vollers. *Ice Bound: A Doctor's Incredible Battle for Survival at the South Pole*. New York: Hyperion, 2001.

Wheeler, Sara. *Terra Incognita, Travels in Antarctica*. New York: The Modern Library, 1999.

Polar Exploration and History

Alexander, Caroline. *The Endurance: Shackleton's Legendary Antarctic Expedition.* New York: Knopf, 1998.

Bancroft, Ann, and Liv Arnesen. *No Horizon Is So Far: A Historic Journey Across Antarctica.* Cambridge, MA: Da Capo Press, 2003.

Behrendt, John C. *Innocents on the Ice: A Memoir of Antarctic Exploration, 1957.* Niwot, CO: University Press of Colorado, 1998.

Bryant, John H., and Harold N. Cones. *Dangerous Crossings: The First Modern Polar Expedition 1925.* Annapolis, MD: Naval Institute Press, 2000.

Cherry-Garrard, Apsley. *The Worst Journey in the World.* New York: Carroll & Graf, 1989.

Cook, Frederick A. *Through the First Antarctic Night: 1898–1899.* (First published 1909.) Reprint: Pittsburgh: Polar Publishing Co., 1998.

Doumani, George A. *The Frigid Mistress: Life and Exploration in Antarctica.* Baltimore: Noble House, 1999.

Fleming, Fergus. *Ninety Degrees North: The Quest for the North Pole.* New York: Grove Press, 2001.

Gurney, Alan. *The Race to the White Continent.* New York: W.W. Norton & Company, 2000.

Huntford, Roland. *The Last Place on Earth: Scott and Amundsen's Race to the South Pole.* New York: The Modern Library, 1999.

Leary, William M., and Leonard A. LeSchack. *Project Coldfeet: Secret Mission to a Soviet Ice Station.* Annapolis, MD: Naval Institute Press, 1996.

Neider, Charles. *Antarctica: Firsthand Accounts of Exploration and Endurance.* New York: Cooper Square Press, 2000.

Officer, Charles, and Jake Page. *A Fabulous Kingdom: The Exploration of the Arctic.* Oxford: Oxford University Press, 2001.

Preston, Diana. *A First Rate Tragedy: Robert Falcon Scott and the Race to the South Pole.* Boston: Houghton Mifflin, 1998.

Reader's Digest, Australia. *Antarctica: The Extraordinary History of Man's Conquest of the Frozen Continent.* Sydney, Australia: Reader's Digest, Australia, 1998.

Solomon, Susan. *The Coldest March: Scott's Fatal Antarctic Expedition*. New Haven: Yale University Press, 2001.

Stark, Peter. *Ring of Ice: True Tales of Adventure, Exploration, and Arctic Life*. New York: The Lyons Press, 2000.

Vaughan, Norman D., with Cecil B. Murphey. *With Byrd at the Bottom of the World: The South Pole Expedition of 1928–1930*. Harrisburg, PA: Stackpole Books, 1990.

Polar Science

Ainley, David. *The Adelie Penguin: Bellwether of Climate Change*. New York: Columbia University Press, 2002.

Alley, Richard B. *The Two-Mile Time Machine: Ice Core, Abrupt Climate Change, and Our Future*. Princeton: Princeton University Press, 2000.

Chester, Jonathan. *The World of the Penguin*. San Francisco: Sierra Club Books, 1996.

Fogg, G. E. *The Biology of Polar Habitats*. Oxford: Oxford University Press, 1998.

Green, Bill. *Water, Ice and Stone: Science and Memory on the Antarctic Lakes*. New York, Harmony Books, 1995.

Long, John. *Mountains of Madness: A Scientist's Odyssey in Antarctica*. Washington: Joseph Henry Press, 1957.

Mayewski, Paul Andrew, and Frank White. *The Ice Chronicles: The Quest to Understand Global Climate Change*. Lebanon, NH: University Press of New England, 2002.

Naveen, Ron. *Waiting to Fly: My Escapades with the Penguins of Antarctica*. New York: William Marrow and Company, 1999.

Polar Travel

Rubin, Jeff. *Lonely Planet: Antarctica*. Melbourne, Australia: Lonely Planet Publications, 2000.

Stonehouse, Bernard. *The Last Continent: Discovering Antarctica*. Norfolk, England: SCP Books, Ltd., 2000.

Swaney, Deanna. *Lonely Planet: The Arctic*. Melbourne, Australia: Lonely Planet Publications, 1999.

Websites

The following websites are just a small taste of the thousands you can find on various aspects of the polar regions.

Antarctic Jobs and Research Grants

www.nsf.gov/od/opp/opportun.htm
National Science Foundation information for scientists interested in conducting Antarctic research

www.polar.org/hr/employ/index.asp
Raytheon Polar Services Employment Information

tea.rice.edu
Teachers Experiencing Antarctica and the Arctic

www.dmna.state.ny.us/ang/109.html
New York Air National Guard, 109th Airlift Wing

www.gocoastguard.com
U.S. Coast Guard Job website (around Antarctica and in the Arctic Ocean)

Antarctic Scientific Research

nai.arc.nasa.gov
NASA: Astrobiology Institute

www.resa.net/nasa/xlife_intro.htm
NASA: Life on Other Planets

www.psrd.hawaii.edu/Feb02/meteoriteSearch.html
Planetary Science Research Discoveries: Searching Antarctic Ice for Meteorites

www.secretsoftheice.org/expedition/index.html
Secrets of the Ice Antarctic Expedition

Antarctic Treaty

www.state.gov/g/oes/rls/rpts/ant
U.S. State Department, Antarctic Treaty Handbook

Arctic Science

www.sfos.uaf.edu/basc
Barrow Arctic Science Consortium

Climate and Environmental Issues

arcticcircle.uconn.edu/VirtualClassroom/anwrref-vc.html
The Arctic Circle website

www.ccamlr.org
Convention on the Conservation of Antarctic Marine Living Resources

www.ipcc.ch
Intergovernmental Panel on Climate Change website

News from the Arctic and Antarctic

www.polar.org/antsun/Sun012603/index.html
The Antarctic Sun

www.antarctican.com
The Antarctican

www.usatoday.com/weather/resources/coldscience/acoldsci.htm
USATODAY.com Science in the Earth's Cold Regions

Polar Webcams

psc.apl.washington.edu/northpole/LatestPhoto.html
The North Pole webcam

www.cmdl.noaa.gov/obop/spo/livecamera.html
The South Pole webcam

Science Organizations

www.comnap.aq
Council of Managers of National Antarctic Programs

www.nsf.gov/od/opp/start.htm
National Science Foundation Office of Polar Programs

nsidc.org
National Snow and Ice Data Center

www.spri.cam.ac.uk
Scott Polar Research Institute

Travel in the Arctic and Antarctic

www.expeditionnews.com
Expedition News: A monthly review of significant expeditions, research projects, and newsworthy adventures in the polar regions and elsewhere

www.explorenorth.com
Explore North

www.arctictravel.com
The Nunavut Handbook

www.greenland.com
Greenland travel information

www.svalbard.net
Svalbard Tourism Office

www.iaato.org
International Association of Antarctica Tour Operators

Index

Symbols

A

B

C

D

G

H

I

P–Q

R

S

T